SERVING THE PIECES

History and Tales
From
The 242nd Field Artillery Battalion
World War II
1941 – 1945
United States and Europe

ISBN 978-0-7414-3240-7

Published by:

INFINITY PUBLISHING.COM

1094 New DeHaven Street, Suite 100
West Conshohocken, PA 19428-2713
Info@buybooksontheweb.com
www.buybooksontheweb.com
Toll-free (877) BUY BOOK
Local Phone (610) 941-9999
Fax (610) 941-9959

Printed in the United States of America

Published January 2013

This book is dedicated to the memory of those who lost their lives serving with the 242nd Field Artillery Battalion in World War II among whom were the following I counted as good friends:

Sgt. Raymond Pfander – San Francisco, California

PFC Marvin Woodworth - Shawnee, Oklahoma

PFC Dale Roberts – Clearfield, Utah

PFC Naiphe – Tulsa, Oklahoma

Lt. Allen Burton – Chester, Connecticut

The following is an account of my experiences during World War II while serving with the 242nd Field Artillery Battalion, from training in the United States to combat in the European Theater. - Edward Walsh, Seward, NE 1993

In 2000 my wife Jeanne and I completed a web site http://www.fortunecity.com/victorian/essex/1/merry.html about the 242nd Field Artillery Battalion.

In 2004 my grandson Vince Walsh completed a second web site http://www.242nd-field-artillery.com/. While in training and in combat I took several hundred photos, which with the help of Jeanne and family I have now catalogued. These photos with accompanying anecdotes and history are now available on CD for those interested in a visual history of the 242nd FA Battalion. - Edward Walsh, Seward, NE 2006

Lower Left – Third Army Patch – Worn on the right shoulder

Lower Right – Seventh Army Patch – Worn on the left shoulder

Bottom – The 242nd Emblem – Appeared on lapel buttons. The cannon was the symbol for field artillery. The beaver head signifies the unit was activated in the "Beaver" state of Oregon. Serve the Pieces Faithfully was the motto of the 242nd FA Battalion.

CHAPTER ONE

Received notice in July of 1942 from the Selective Service to report to the Induction Center at Fort Crook which was old Fort Omaha at Omaha Nebraska, on August 7, 1942. There were probably 30 or 40 men who boarded the train early that morning at the old Burlington Depot in Seward, Nebraska for Lincoln, Nebraska and then on to Fort Crook.

The Army medical tests were embarrassing for nearly everyone. We had to take off all of our clothes, even shoes, and go room – to - room doctor – to - doctor. There was one room for hearing test, one room for weigh-ins, one room for measurements (how tall, how long your arms were, how long your legs were, normal chest measurement, chest expanded measurement, how long your fingers were, how far around your upper arm relaxed and expanded. How long were your feet, how far around your stomach and hips, and if you had ten fingers and ten toes. (They only had one thing left to measure, but they didn't do that – then. They looked that over later!)

There was a room for hearing tests and one to check for hernias and hemorrhoids, a room for checking teeth for cavities, inlays or false teeth. A room for checking heart, pulse and temperature and other rooms I don't remember!

It was like an assembly line in a factory. You had to stand and wait between rooms until they were done with the ones ahead of you. All the time completely bare! This took at least four hours or maybe five. It was great to get your clothes back on again!

We took tests and physical exams all day and those of us that passed were sworn in to the US Army just as the sun was going down. We were in the Service and drawing pay! We would have two weeks at home before going to a Reception Center.

This was the first month that the Army had raised pay for a private to $50.00 a month. It had been $21.00 a month for years and years before this.

August 21, 1942 --- Boarded the train about 7 PM at the Seward Depot, there were about 21 or so Seward County guys that had passed physicals on August 7. The train already had many inductees on it. The train stopped in Utica a few minutes and the local liquor store man, his name was Groce, brought down a bottle for me and one for Lester Hans (Utica, NE), 2 fifths, we just passed it around and it was soon gone.

First thing on the train out of Seward, there was a crap game going with a bunch of black guys and even though I was 22 years old, I had never heard language like they used before!

We picked up more guys at York and Grand Island and then changed trains to the Challenger, or a streamliner. We picked up more men through the night in western Nebraska, probably North Platte, maybe Sidney, then into Colorado and arrived in Denver late afternoon of August 22, 1942.

We left the trains and got on Army trucks and went to Fort Logan, a Reception Center. It was raining when we got off the train. We were fed and showed a place to sleep. Got to bed about 11 PM, was dead tired.

Got up at 5 AM, ate and took a hike down and back up a hill or mountain, some could hardly make it, were not in shape, and the air was different at high altitude. Then came shots and inoculations and speeches and films and tests, and dinner, more films and tests and last thing the Army Ground Forces test or I.Q. test, always given after you are tired and worn out and under stress.

Then a haircut, clear to the skin, and complete equipment issued: clothes, shoes, shine kit, A & B duffle bags, canteen, belt, musette bag and harness, dress clothes and fatigue clothes, dog tags and told to memorize the serial number on the tags: Army serial number 37452619. If the number starts with 3, it means drafted. If it starts with a zero, it means enlisted. The second number means corps area. Seventh Corps was Nebraska and several other surrounding states.

Here is where everyone found out about open showers and toilet stools. This place must have had 30 of each. Most of these farm boys had never had indoor plumbing at home. Maybe some had been in open showers in high school but not many had. Some were really embarrassed but I thought, what the hell, everyone is naked again, just like the medical at Fort Crook!

Seward, Nebraska Train Depot, 6th & Ash Street

August 21, 1942

Draftees preparing to depart

Draft Board Members back row left

Edward Walsh, Gresham, Nebraska back row

CHAPTER TWO

August 14, 1942 ---- Next morning back to the train and heading west again.

Crossed the Rocky Mountains, and then down to Ogden, Utah. We got to leave the Troop train for several hours and looked the town over, then the train crossed the Great Salt Lake or at least a part of the marshes, and on to Sacramento, California. Again we had several hours to look over the capitol of California, the parks, etc.

Back on the train and headed north. We still did not know to where. After almost two days we were at Medford, Oregon, a fair sized town just above the California border. There we left the train and got on Army trucks, six by six. We were to find out later they were so called because they could have six wheels on each side by making the front ones duals and all six on each side had power to them.

August 27, 1942 ---- A few miles out of Medford, we came to Camp White, a basic training camp for artillery units and also the home of the Ninety-first Infantry Division. This area in the Pacific Northwest is the 9th Corps area, headquarters at Fort Lewis, Washington just outside of Seattle. Camp White is located on the Agate Desert, about a half-mile from the beautiful, wild and rocky Rogue River. The valley where the camp is located is flat and covered mostly with small gravel type rock (agates).

Right in the area of the camp it is almost impossible to dig a hole. After about an inch of gravel there is a couple inches of volcanic dust and then below this is black basalt embedded with gravel or agates spaced about every two or three inches as far as you want to dig. We used to practice digging fox holes and slit trenches with a shovel and pick. You could not dig a hole big enough to hide in or to have a tank run over you in two days, it was that hard and sticky. However, up and down the valley off camp where this soil is planted to hay or grass and irrigated, it was as green as the hills of Ireland. Mountains rose above this valley on both sides but were not of great height. In this area the mountains were covered with pine trees.

Through the Rogue Valley and the small valleys coming into it there were small farms which were mostly planted to fruit trees, apple, mainly, but these farms had been bought up for miles around by the government to make up the Military reservation of Camp White for maneuvers, artillery firing range, rifle ranges, obstacle courses and so forth.

Far to the east you could see several snow capped mountains. I believe one of them was Mount Pitt. Also Crater Lake was over that way. It was a mountain that had a volcano and later filled with water making a beautiful lake on top of the mountain. This whole area was once volcanic and the debris and dust from them covered the soil in great depths in most places.

We were to be part of the 242 Field Artillery Battalion. The Battalion had been activated on August 20, 1942. Major Walker R. Goodrich assumed Command. On the same day the Cadre arrived, the non-commissioned officers. Master Sergeant or Sergeant Major, five first sergeants, staff sergeants, buck sergeants and corporals and 5 officers to make a total of 88 men. These men came from the 76th Field Artillery. Other officers arrived by August 24 from the IX Corps, fifteen being Second Lieutenants direct from Officers Candidate School (OCS) at Fort Sill, Oklahoma. Then the fillers, like us, arrived from August 27 to September 3. Others came from Camp Roberts Replacement Center in California. These men had already had 13 weeks basic training so they were part of the nucleus that already knew a little about artillery. The rest came from Reception Centers at Fort Leavenworth, Kansas, Fort Logan, Colorado, Fort Snelling, Minnesota, and Headquarters Service Command, Fort Lewis, Washington.

Major Goodrich, the Commanding Officer was a West Pointer, probably the oldest man in the Battalion. He had been in the Service a long time. Smart, friendly, easy going and knew his job. He had a terrific memory for names and faces, destined to move on to bigger things.

Second in command Major John F. Brownlow, Knoxville, Tennessee, was small, thin, wiry, and bowlegged, a Southerner, and also a West Pointer. He was tough, sarcastic, loud mouthed and out to better his rank; younger than Goodrich but a direct opposite as far as personality, he also knew artillery, maybe even better than Goodrich. He had white hair, not blond, white. His chin was set back quite a bit to the point where he looked like he was always grinning, and when he walked on his little bowed legs (which really stood out because everyone wore leggings) he leaned forward as if going into the wind.

(I think now that this man must have patterned himself after, or wanted to be like, some of the famous people in military history, which I'm sure he had studied in his years at West Point, perhaps Phil Sheridan, or George Custer, John Reynolds or even Southern Jubal Early or Jeb Stuart of Civil War times.)

In October Major Goodrich was promoted and it was said that he went to Hawaii. I don't know if this was true. Major Brownlow then took command and remained in command for the duration of the war. He was promoted to Lieutenant Colonel and toward the end of the war to full Colonel, (temporary).

So we arrived at Camp White and were assigned to a battery. In the Artillery, the group of men who serve a cannon is called a "battery". There were 5 batteries, Headquarters & Headquarters Battery, A, B, C, and Service Batteries. Within each battery are "sections" according to the responsibilities they have. I went to A (Abel) Battery, but no one was assigned to a section until we had our 13 weeks basic training, except we did all have our own sergeants like a section.

We were fed and assigned a bed and went to the supply room to draw our blankets, sheets, and pillow. We had to sign for these so we would turn in the same amount for laundry. We also signed for a footlocker (much like a small trunk with a tray in the top).

The barracks were new, made out of wood, regular wood siding on the outside but the walls were not, and never would be finished on the inside, just 2x4's and rafters and supporting beams run from the ground up. Everything smelled new, like new wood, inside the barracks. Outside it smelled like fresh oiled road because all the streets were new asphalt. The water was milky colored and tasted of new pipes.

Everyone slept in the two big rooms, one room down stairs, and one room upstairs, except the 4 highest-ranking sergeants who had rooms, 2 in a room. The bathroom had about ten showerheads, and there were about as many stools and sinks.

Right next door was the kitchen and mess hall and to the back a building that had 2 offices in it, one for the Captain and one for the First Sergeant and the battery clerk or record keeper and the mail clerk. These last two were both corporals.

There was also a Medical Section that was separate from the batteries. It was headed by a Captain who was an MD, a couple of tech sergeants, two or three tech corporals and 2 or 3 privates.

The above building also had a Supply Room where they had clothing, blankets, rifles, pistols, canteens, belts, musette bags, etc. that were issued to the troops that didn't have them or needed them replaced. The Supply Sergeant was responsible for all equipment and had a record of everything, signing it out to whoever needed it, the signer then being responsible for it. Anything from handkerchief to a big truck or Howitzer, and "if you lose it, you pay for it", like the driver of a truck or a jeep, the sergeant at the head of a gun section for a 105mm howitzer, a pistol, rifle, belt or anything!

The far end of this building was called a day room. It had chairs and tables, books, pop machine, and any recreation thing that could be rounded up (very few at this time in 1942.) This room was also used for classes and meetings.

The camp was set up in squares like city blocks only the squares were much larger, I would guess about ¼ to 3/8 mile square. The roads were black top, ground gravel of pebbles mixed with clay and basalt. Paths between buildings were larger rocks brought in from the mountain areas and then graveled. The buildings were painted a dull tan.

Just to the right of this was the motor pool and shops or garages with all the trucks, jeeps, and trailers lined up in front of the shops on a big blacktopped area. Each battery had about 6 or 7 big 6x6 trucks or prime movers they were called. Four pulled howitzers, one the fifth section with mounted machine guns, 50 cal., one the field kitchen and cooks. The Captain had a command car and the other officers had Jeeps with drivers and other riders. Wire and survey sections had four wheel drive pickup type vehicles 4x4's with radios, spools of wire, telephones, survey equipment, etc.

The first night in Camp White, my bed was in Sergeant MacFarland's (Minnesota) section and the bed next to me was an old guy that had been in World War One. He had volunteered for World War II (which he did not make, was finally let out for medical reasons, or being over 35 years old.) Anyway, he said, "They are going to get us up about 4:30 or 5 in the morning and we'll have to be dressed and outside for roll call in about 15 minutes. About two thirds of these guys won't be dressed right and we'll get chewed out. So I'll tell you what to put on and how to button things up and especially the leggings and to blouse the pants legs over them."

So he showed me how to do the leggings that night (the right way to lace them) and he said, "I'll bet you will be the only one of these new guys that come in today that don't get chewed out!"

So at 5 AM there were whistles blowing all over the place and Sergeants yelling, "Off your ass and on your feet, this is Reveille and not Retreat." "Roll call in 15 minutes." And this was as wild a mess as you ever seen! Guys trying to get dressed and most of them did not make it. The whistles were blowing again and Sergeants yelling: "Fall out!" and "Line up for roll call!"

Some did not have both shoes on or laced. Some had one legging laced wrong. Some had no cap. Some with shirts not buttoned or buttoned crooked. The first Sergeant and the other Sergeants tried to get men lined up in two rows at equal intervals (dress right dress). What a shuffle! Finally the first Sergeant started calling the roll alphabetical and griping because the answer (here or yo) wasn't loud enough. Finally after the Z's, he does an about face and says, "All present and accounted for, sir!" to the Captain.

The Captain says, "The rest of the lower officers will inspect the troop for the dress code of the day." They went down the rows and bawled out every one of the troops that were new except me! They even got on some of the ones that were from Camp Roberts that already had 13 weeks basic training. The old World War I guy, his name was Ransom, really got a bang out of this. He was a great help to me those first few days!

Then we had about 30 or 40 minutes to make beds, wash, shave, and it was time for breakfast at the mess hall, cafeteria style. We had chipped beef and gravy on toast, cold cereal, dried fruit and coffee.

Then back to our bunks for a half hour or so and then fall out and we were marched out on parade grounds where other batteries were also and the whole battalion had 50 minutes of calisthenics. This was really a mess for many days - - - getting everyone to do them by the numbers in cadence. Some guys were never able to do some things, especially pushups. I was in pretty good shape and could do it all except for the total amount of pushups. They wanted, I believe, it was 50. The most I or many of the guys could ever do was around 20. Probably only 10 or so of the 500 men could do 60. I know now they were probably old military men or athletes. I remember 2 of them were young small college coaches. We did pushups, side-straddle hop, Philippine squat deep knee bends, and many coordination type things, finger, arm, leg, neck exercises, running in place, bicycling your legs laying on your back and many others.

Then we would have classes for 50 minutes (no smoking), and then a 10 minute break for smoking, rest room, etc. We would have 50 minutes on learning general orders that every soldier who does guard duty must know. Then 50 minutes on first aid, 50 minutes on close order drill, manual of arms with rifle, cleaning and assembly of rifle, memorize brand of rifle, model number and names of parts. (Eventually the assembly of rifle had to be done blindfolded!) All were 50 minute classes.

We learned to do 30 cal. Springfield rifles, 30 cal. Enfield British rifles, 30 cal. carbines, short barrel M1 infantry rifle, 45 cal. Colt automatic pistol, 30 cal. and 34 cal. machine pistol, 30 cal. machine gun water cooled, 22 cal. heavy barrel target rifles, 50 cal. air cooled machine guns, also the Bazooka antitank weapon rocket type recoilless M I with armor piercing rockets.

The line on the 50 caliber went about like this: "This is a 50 cal. Machine gun Model M One – H.B., flexible ground mount. It is belt fed, heavy barrel, air cooled and gas operated. It fires 500 rounds per minute of regular 50 cal. cartridges or 50 cal. tracer bullets or armor piercing bullets. It has an extra barrel to be changed during rapid firing when other barrel gets hot. Barrel must be put in and backed off so much so it had head space for heat expansion. The cartridges are set up in chain link belts and stored in metal boxes. It is loaded by putting front cartridges of belt next to bolt and pulling back and releasing bolt to put first cartridge into firing chamber."

All of the above was done over and over until we memorized names, parts, etc, and the guns could be assembled and reassembled blindfolded!

We would have 50 minutes of chemical warfare, poisons, how used and how decontaminated, how to breathe gases, the burns from the different kinds, gas mask drills, how to use, etc. Fifty minutes on truck maintenance, from greasing to washing, but nothing major. (The trucks, jeeps, etc., went to Ordinance battalion for any major repair or anything more than regular maintenance.) Fifty minutes on military customs, discipline, saluting, emblems, rank, outfits, uniform, size of different groups, like squad, platoon, section, company, battalion, regiment, groups, division, corps, armies, etc. Fifty minutes of routine, from getting up in the morning until going to bed at night and even at night for the ones on guard duty.

We had 50 minutes twice a day on the 105 millimeter Howitzer and the nomenclature of it, the shells and how to operate, load, and fire the cannon, and the practice routine of doing this called the "Cannoneer's Hop".

Seven men, a gunner, assistant gunner, two second assistants and three cannoneers serviced each Howitzer. The rest of the (gun) section was the driver of the 6x6 truck that pulls the Howitzer; one of the cannoneers was the assistant driver in case the driver gets killed. The section is run by a sergeant. This makes 9 men in a section, with 4 sections in a battery, and 3 batteries in a battalion.

The 36 men of the battery are commanded by one First Lieutenant, who is the executive officer, and one Second Lieutenant who is the assistant executive officer. There is one telephone operator, one radio operator and one recorder, a man who keeps track of the Howitzer on paper for deflection, elevation, alignment, type of shell (high explosive, white phosphorus, armor piercing), type of fuse (quick - blows up when it hits; delay - blows up after it penetrates; or time, which can be set to blow at any elevation) each time the gun is fired. Also the number of powder bags up to 7 in the shell case for range needed. These bags were numbered and varied in size. The range of the Howitzer is up to maybe 8 miles, but with great accuracy up to 5 or 6 miles. The recorder keeps track of shells expended.

There were very few guys that could handle this recorder job even if it was just simple arithmetic, mostly adding and subtracting because they had to be fast. In Fort Sill when we were firing for Officers Candidate School or FAS (Field Artillery School), we would fire from a few up to 2000 rounds (shells) a day. I did recording day after day for a battery in Fort Sill until I was transferred to the Survey Section as a messenger or agent.

There was a guy from Chicago by the name of Lepinsky that did quite a bit of it after me. After Fort Sill, he got out on a medical discharge, or was too old. This was in Camp Howze, Texas, probably after we passed the Army Ground Forces Tests to go overseas. I don't know who was trained to do it in combat and I don't remember seeing who was doing it.

So now we have about 41 men serving these 4 Howitzers.

The 42nd and 43rd men are the Howitzer mechanic and his assistant. They do all the maintenance on the Howitzers and also teach the 36 in the gun sections how to do it.

The next section is the 5th section - machine gun section - that does the guarding on the far perimeter around the Howitzers. Other guys are walking guard with rifles. Also the fifth section hauls all the extra ammunition, about five or six truckloads in addition to what the gun sections haul.

Other 50 minute classes were held on hygiene, social disease, etc., with films. Fifty minutes of films, aircraft identification, British, German, Italian, U.S., Russian, Japanese, also uniforms, ships, tanks, guns. Fifty minutes of films on actual European theater of war and the Pacific theater. Fifty minutes of talks, orientations, etc. There were these and others that went on over and over, especially during the 13 weeks of basic training.

We were told that there would be no passes to go off the base until the 13 weeks of basic were over, but after 5 or 6 weeks, they started putting out passes to go to town in the evening from 6 to 12, also from Saturday noon until Sunday night at 11 PM. Only 30 to 50% of the personnel could be gone at one time.

From the beginning you could go to the Post Exchange PX in the evenings or Saturday afternoon and Sunday. Also there were several theaters in camp where shows changed 2 or 3 times a week. One theater was only 2 blocks from our barracks. After we had seen the show near us we would walk the odd miles to another theater and see the show there. We even walked about 4 miles across the camp to the big main PX one evening. Four miles over and 4 miles back, quite a hike.

One thing I remember about the theaters was the smell of mothballs. It was the rainy season in Oregon, rain, mist or fog, and even wet snow almost every day. It was cold so everyone wore the heavy wool overcoats and uniforms that had been issued. These had been stored in moth balls in the warehouses. When they got wet walking to the theater, it brought out the smell of moth balls and with about 500 guys in the theater in their wet overcoats, it smelled like the place was being fumigated!

Most of us went to the PX or the movies every night unless we had guard duty or KP.

When we would go on pass into Medford, Oregon, like on a Saturday afternoon, we would get a hotel room and then go do some sightseeing, also go into the stores and bars – drink some beer, then go to a café and eat a steak and back to some of the bars and USO clubs, then back to hotel at midnight and sleep until noon on Sunday, then eat breakfast. Some guys would go to the USO where some family would take them home for dinner and sightseeing around the country or town in the afternoon. I never did this. I usually ate dinner, went to a matinee show and then back to camp by 5 PM on Sunday in time to eat supper at camp.

Some guys went to Grand's Pass or Klamath Falls, Oregon or by bus down into California to Yreka. I went to Yreka one weekend but spent much of the time riding on the bus there and back. Not much fun, because it went through such high mountain passes that it made my ears hurt!

The bars in Oregon at that time were a little like the bars shown in old western movies. More of a saloon than a bar. The back quarter of the buildings were separated from the front by lattice work dividers and in the back were legalized card playing tables, poker, blackjack, etc.

Also the patrons were the western type seen in movies. There were sheepherders, cowboys, miners, and a few farmers and fruit growers. But most were lumber-men, rough, tough looking people, wearing wool mackinaws, wool plaid shirts, cowboy boots, mining boots, leather coats, even buckskin pants and shirts, beards and long hair (even in these years of 1941 and 1942).

Ginger Rogers, the movie star, had a ranch or a home near Medford. One Saturday night she was in Medford selling War Bonds. They had a big stage out in the middle of main street and she spent about 2 hours singing, dancing, talking, and selling bonds. The main street was packed with local people, soldiers, stockmen, miners, lumbermen, and everyone really bought the bonds!

I was surprised at how Ginger Rogers looked. In the movies she is blond, but her hair was light red and her face and arms were solid freckles. They said that when she was not making a movie she spent her time at home near Medford, Oregon.

Edward Walsh, Gresham, Nebraska

Photo Booth, Medford, Oregon

Fall 1942

CHAPTER THREE

The first six weeks or so in Camp White, the weather was blistering hot in the daytime on the Agate Desert, but the nights were cool. Everyone went without a shirt in the daytime and really had a tan. The general that had the 9th Corps area was an outdoor nut and physical fitness nut. So this brought on the "no shirts" long hikes with full field pack 90 pounds and all kinds of exercises!

There was talk that everyone had to cross the Rogue River with a full pack before basic training was over. I worried about this as others did because I could not swim, or even float. Many others could not swim but this was small comfort because I thought they would have their hands full with 90 pounds of pack even if they could swim. There were even rumors that some of the infantry of the 91st Division, the Pinetree Division, had drowned doing this.

I knew later that the way the river was crossed was that several good swimmers would cross the river with ropes and stretch them from side to side and the rest would cross holding on to the ropes. The Rogue was not very deep in this area but it is a wild river and has terrific speed running over the rock bottom. It would be hard to keep from drifting downstream.

It started raining after the first six weeks of sunshine and it rained every day just a gentle steady rain, sometimes mixed with snow. You could see the clouds coming in because they were lower than the mountain tops and it was wet, wet, wet, and cold. There did not seem to be much sickness from it although I was sick one time and had a fever so bad I did not know what was going on. They took me to the battalion doctor and I was sent on to the camp hospital. Was there I don't know for how long now and I don't believe I did then. Finally got okay and went back to the outfit.

On payday, the last day of the month, everyone was paid in cash. This was the day the American Eagle (U.S.) paid the GIs or commonly called in the Army, "the day the Eagle shits."

The pay officers of the battalion take a driver and 2 guards and go to Camp Headquarters to pick up the payroll for 500 men. They bring it back to Battalion Headquarters and the battery pay officer and 2 guards go to Headquarters and get the payroll for the battery of about 100 men. They bring the payroll back and put the money in stacks on the table in the day room or mess hall. The pay officer plus a witness officer and the company clerk with the payroll book sat at the table with two guards behind them. The men line up by highest rank alphabetically down to the last private. When your name is called, you salute the officer and the clerk says how much your pay is. The officer counts it out, and then you count it back to him and sign your name in the payroll book, salute again, and take the money. (At this time a private got $50.00 a month.)

Behind the pay line was always a collection line, guys collecting from ones who owed them before they had a chance to get into a card game and blow their pay before they paid off their debts! Sometimes it took all the money some were paid to pay debts and they would start trying to borrow on next month's pay. Or they would lose all their pay the first night at cards or dice trying to make a bundle!

That night the card games and the dice games started. Of course, two or three usually wound up with all of the money from the guys that were the gambling type. The money others borrowed from them usually had high interest on it, like 10% or 25% for one month. I also lent some this way but you had to be careful to whom because they might not pay it back, or they might get transferred!

I played some poker and blackjack, but never the dice. Usually won a little or at least broke even in the small games here in the states. Got in a few large ones in England using British money, was several hundred ahead at times but when we crossed the Channel to France I was down to about a little over two hundred, which I sent home.

After lights out at 9 PM on payday, the dice and card games went on in the showers and toilets all night with someone keeping an eye out for the officer of the day and guards, putting the money away if they came around. Of course, the guards knew what was going on in every shower in every camp all over the world. They just ignored it, - making Army tradition - one of those don'ts that everybody does, everyone knows it, and no one does anything to cause any problems. Just look the other way for a few nights, and a few men will have all the money and will be lending it out all month to the ones that are broke, a little at a time at high interest so they have money to buy cigarettes, soap, toothpaste, a beer or go to a movie once in a while. But they never learn. If they can borrow even two or three dollars they will want to get a game started to see if they will have better luck.

Also after payday there was quite a lot of drinking for a night or two. One of the wildest things that happened was one night after payday, an older guy, an Indian that everyone called "Chief", got drunk and came back with a bottle and got to singing Indian songs and doing dances so finally some of the guys took him down and put him in the shower so the rest of us could sleep. He was discharged later because he was over 36 years old. They thought he really didn't know how old he was and his service record did not show it either but he looked like he was over fifty. He had volunteered to get in the Army.

We practiced day in and day out on foot drill, cannoneer's hop, manual of arms with rifles, machine gun practice, practice at digging fox holes and slit trenches, guard mounts and all the other things.

Eulogio Yanez (Los Angeles, CA) was the Motor Pool Sergeant. He came from the barrio in Los Angeles and was in the original cadre that started the 242nd. The cadre was formed out of the 76th Field Artillery Battalion, which had just recently changed from horse-drawn to mechanized. They had been stationed at Fort Ord, California, and had done shore patrol at the Presidio of Monterey immediately after Pearl Harbor with horse-drawn 75mm cannon.

When the switch was made to trucks, each battery had to have a sergeant who was a truck mechanic so guys like Yanez were brought in. He was a good mechanic and received three months training at the Field Artillery School in Fort Sill, OK to learn the Army way of doing things.

Yanez had a Mexican accent which was strange to us Nebraska born, and he was always laughing and telling jokes. He knew the background of everyone in the battery because he made it his business to find out. He went into our service records and learned where we were from. He called us sheepherders, cowboys, plowboys, coal diggers or whatever the background was.

Anything on the trucks like the bell housings on the front wheel drives had to be kept clean and oiled so they would not rust. He would say, "Now you plow jockeys would not let your plow rust, would you, so let's have that bell housing clean and oiled." If someone would ask, can I do this or that this way? He might say, "Whatever gets your gun off, plow jockey." Everyone called him Poncho except when the officers were around then it was "Sergeant Yanez!"

He called the Mexican boys in the outfit "black boys". He used expressions like "Grandma is slow but she is old so speed it up." Or "You're going about that like old people having intercourse." He had lots of jokes about "Kilroy" and used the word forms SNAFU (situation normal all fouled up) or TARFU (things are really fouled up). "Shit house mouse went the wrong way".

In 1990 someone in the old outfit went to see him in Los Angeles. Said he was still a comic, maybe better than 50 years ago. If that's possible!

When we were first in Camp White, a couple of us were walking up the street after we were off duty in the evening and we met someone coming toward us. As we met, he stopped us and bawled us out. He said, "Don't you salute officers when you meet them?" We said we did not know he was an officer because no one wore shirts. So he said we could tell he was an officer because of the boots and pants he wore and the pith helmet he had on, but he guessed we were right that it was hard to tell and officers should put their insignia on the pith helmet and that he would see what he could do about it. He said he was a captain and a West Pointer. Anyway, in a few days, the officers were wearing their insignia on their helmets!

A few days later, I got put on report by First Lt. William Winchester (San Francisco, California) because I asked, "WHY?" I should do something or other. He said, "You don't say why to an officer. You just do it!"

Anyway I went to the Battery Commander's office and told the First Sergeant I was supposed to report to the Battery Commander. (This was one of the 50 minute classes we had had on how to report to the Battery Commander.) The First Sergeant said okay, so I knocked on the Battery Captain's door, he said, "Come in." I walked up within about 3 feet of his desk, stood at attention and saluted. He saluted back and I said,

"Reporting to the Battery Commander as ordered, sir." He said the Lieutenant had told him why I had been put on report and what did I have to say. I said I didn't understand what he meant. He said, "That's not it! You can ask him why, but you have to say 'Why, sir?' If you can remember that you'll be okay." He kinda grinned, and said, "Do you have any questions?" I said, "No, SIR." and he said, "Dismissed." I saluted and he saluted and I did an about face and walked out.

As I went through the First Sergeant's office to leave, the sergeant followed me out and said, "Just a minute, kid, I want to talk to you." He said he thought the Lieutenant was pretty picky doing this and that he would see to it that the clerk did not put it on my service record. He also said that the Captain had told the Lieutenant not to do this again as he had pulled this on several new recruits and that the Lieutenant was just supposed to tell the recruit to say, "Sir" and not to put them on report. No one had gotten any extra duty or marks on their service record for this.

The First Sergeant told me, "You can say most anything to an officer as long as you say 'Sir'". He had been in the Army for over 20 years and he sure could! These old career army men knew every angle and could really read out the officers and there was nothing these "90 day wonders" could do about it.

We had an old Army sergeant (over 20 years) that was the foot drill sergeant. Many of us had trouble keeping in step. One night he called 10 or 12 guys out after hours (after supper) to practice foot drill. I was one of them. He did this on his own without asking any officer or any of the section sergeants. His name was George Van Zyl from Finger, Tennessee. He was Dutch and could hardly talk English. Everybody called him "Dew Gee". He wasn't really mean, and all he knew was Army.

Anyway, he got to drilling us out on the parade ground and all of a sudden, here comes my section sergeant, a small, young Irishman with black hair and black eyes, name of MacFarland and he says, "What the hell is going on here, Dew Gee?"

Dew Gee says, "I'm going to show these guys how to keep in step if it takes all night."

MacFarland says, "You dumb Dutch son of a bitch, you can drill some of them but there are two from my section you're not. Walsh and Trussell, get the hell out of there and come back to the barracks with me."

So we went. And then MacFarland got the other sergeants and they went to an officer, and the officer stopped the whole thing.

After a couple of months they told us we would go on a 25 mile hike with full pack but not rifle as rifles were so few we had been using old Springfields and Enfields on the target range. (We had borrowed M1's, the infantry automatic, from the 91st Division Infantry to fire them and learn how to clean and field strip them, reassemble them and learn the nomenclature. The carbine is the artillery man's personal weapon and these were not available yet. All that were being manufactured were being sent overseas.)

Anyway, we started out on the hike, "A" Battery leading the way, as was usually the pattern in everything the next few years. We were strung out for some distance, single file, about 500 men. Most of us were getting in good physical condition and it sounded easy. Everyone walked, officers and all, except the drivers of the 5 field kitchen trucks, a couple of jeeps and the 5 supply trucks, maybe 5 men.

The first 12 or 15 miles were not bad at all. The weather was nice, sunshine, late in the fall, no wind, temperature about 60 degrees. Everything in the country was in fall color. We would walk 50 minutes and rest 10 minutes. Going up some of the mountain trails was kind of hard and the rock underfoot was not so good on the feet.

After 18 or 20 miles the shoulder straps on the backpacks cut into your shoulders pretty bad, so we carried the pack part of the time and then wrapped handkerchiefs under the straps to stop the cutting. Also it was good that the shoes we had been issued were about 2 sizes too long and 2 sizes too wide because your feet got hot and swelled up some. The last five miles were pretty bad and quite a few dropped out and were brought in by the supply trucks.

We finally reached the place where we would camp. Guard posts were set up around the perimeter of each battery and the cooks set up their stoves and started supper. We had had a noon meal off the cook truck also.

Garbage pits were dug and straddle trench latrines (toilets) were dug, (a narrow trench about 6 inches wide, 3 feet deep and 12 feet long so several people could go to the toilet at the same time). This was a new experience to everyone here but the old regular Army guys.

Everyone took turns at guard, 2 hours each through the night. We put our bed rolls on the ground under the trees, got water from large bags called Lister bags hung on tree limbs. We cleaned up and then got in line with mess kits and canteens and went by the cook truck, got our food and coffee, went to our bedrolls. We ate the food and got into our bedrolls and went to sleep. Everyone was really tired but each one got up and did his guard duty sometime during the night. I was lucky and drew an early watch, 7 to 9 in the evening and then I went to bed and slept all night.

It got so cold that night, about 20 degrees, that there was a real heavy frost. The bedrolls were white with frost when we got up at sunup.

We washed and shaved and ate breakfast. Cleaned up mess kits, rolled up the bedrolls and put them in the supply truck they had come out on. We filled the garbage pit and straddle trench, policed up the area of any litter and started the march back to Camp White.

The countryside and mountains were beautiful at sunrise with the frost and a little fog here and there. We were up in the mountains probably 1000 feet higher than the Rogue River valley where the camp was. Everyone was a little stiff and lame from the walk the day before but after a mile or two the stiffness was gone. We moved on at a good pace and walked 50 minutes and rested 10. We went by many small orchards with good looking apples on the trees even this late in the fall. These farms or orchards had been bought up by the government and were now part of the military reservation. We were not allowed to cross fences and pick the fruit because it had not been sprayed or cared for and there were signs up not to eat it because it might cause sickness (such as cramps and runs).

We went on and were back by a little after noon. Got cleaned up and slept until suppertime.

There were some that did not make it through the march and had to be brought back by truck. These were mostly the older men (over 36). Shortly after this, the ones over 36 were released from service nationwide if they wanted out or if they were not physically fit. Most who were not regular Army got out, but a few stayed in that were in good shape and determined in mind. They did okay for the time being, but as a year, and then another, and another came around and we were ready to go overseas, it was getting tougher and tougher for them. More of them had to get an over-age discharge and go home.

As I remember now, we only had 2 or 3 older guys in my battery when we went overseas. Probably more in the battalion, and they did fine until we got into combat and the fall rains and winter snows came.

One day we were out in the mountains practicing reconnaissance survey and occupation of position, digging foxholes and putting out perimeter guards. Several of us were sitting around waiting while the gun crews went through practice dry runs and firing the sub caliber 37mm guns mounted on top of the 105mm cannons because there was no range at Camp White long enough to fire the 105mm shell.

While we were sitting there, old First Sergeant Bert Reed (Witchita, KS) came over and sat down beside me, talked awhile, and then he said, "Kid, (he always called me Kid or Kid Walsh) how would you like to get out of this rain and snow? There is an opening in the kitchen and you can have this job if you want it. You would be corporal in 3 or 4 months and probably a sergeant before the year is over. I know you are used to being outside because of your farm background so it's up to you."

I said I did know how to cook pretty good but believed I would rather stay with the firing sections and thanked him. He said, "That's fine, you'll get to be corporal anyway but it may take a few months more."

We did not have the corporal spots filled at this time in any of the sections. However, there was no chance of being a sergeant in any section except kitchen, unless someone transferred out or was killed. So Corporal was as far as I ever got and I was glad I did not take the kitchen job and so was Sergeant Reed. He asked me about it one time in Europe and I said I sure liked the messenger job I had. He was the guy I picked up morning reports from every morning and delivered to Headquarters. (Morning reports are how many men are present for duty, how many are on sick call, how many in hospital, the inventory of gasoline, 105mm shells, food, etc., also the location of the Howitzers.)

We practiced day in and day out on foot drill, Cannoneer's hop, manual of arms with rifles, machine gun practice, practice at digging fox holes and slit trenches. There were training films almost daily and short speeches by Colonel Brownlow.

He would always talk a few minutes before the films and he always fascinated us Midwesterners because he was a Southerner and a West Pointer. He ended sentences with "You know" or "You see" and he referred to the men as "You people". He hated "Niggers" and said so! He was really "burned" when in England he saw black soldiers out with the English girls. He even gave a talk and said, "Don't go taking out any of these English girls, they all have been wet down by some nigger buck and are not fit for a white man to talk to!" He was really tickled when the 101st Airborne came back from France to rest in our area in South Wales. They got into a fight with the blacks from the Quartermaster Corps at Cardiff, Wales. Before the night was over, a dozen black soldiers were dead, and 87 in the hospital, with 3 whites killed and 30 some whites hospitalized.

I walked my first guard duty at Camp White. I drew the guard post at the motor pool and walked around it for 2 hours, three times in 24 hours. The two hours in the middle of the night were cold and frosty. The sky was clear and full of stars. I could hear the nearby Rogue River roaring and gurgling, and the occasional screech of a mountain lion down by the river or up in the mountains. The two hours at night really wears you down.

I never walked much guard duty stateside because when the roster for guard duty came up every three weeks to a month, the officer of the day, the sergeant of the guard, the corporal of the guard for each post, three privates for each post plus one extra private were on the roster. These men were lined up and inspected by the Officer of the Day. He checked rifles for cleanliness, uniform for neatness, shoes for polish and asked questions about guard duty and what some of the general orders were. The guy that was best dressed with a clean gun and knew the answers, etc., was picked as the Colonel's orderly for the day (or dog-robber as it was called.) This man was excused from guard duty that night and at first call next morning he reported to the battalion commander and stayed in his office or followed him around all day or ran errands for him until retreat was sounded in the evening. A real snap! I managed to have the right answers, the cleanest rifle and neatest dress for this every time but twice while I was a Private or PFC.

It was very interesting to be around headquarters for a day and dog rob for the Colonel. You walked on the left side of a superior officer. It was something to do with ancient times about saber side or something or to protect his side opposite his saber hand. I dog-robbed in Camp White, in Yakima, Washington and in Fort Sill, Oklahoma.

I walked guard only once in Camp White and once in Fort Sill. In Fort Sill I was promoted to Corporal and became Corporal of the Guard when I did guard duty and would have six men under me. I marched them to their posts and then stayed awake in case they called for the Corporal of the Guard if there was a problem. They would call, "Corporal of the Guard, Post #5." The other guards could call the Corporal out to that post to see what was wrong, or maybe wake the sleeping guards to take extra men there.

After Colonel Brownlow took command of the 242nd, he wanted each battery to have a bugler. He wanted them to take turns doing the bugle calls every day at the guardhouse. So he started a school with five guys that knew very little about music. He could not pick people through the army records because nothing was ever asked about musical ability or training at the induction center or the reception center. He had to go for volunteers or just plain order someone to do it.

They started out in Camp White, 2 hours every morning - worse things you ever heard. He borrowed a guy every day from the 91st Division to teach them. This went on all the time at White and they never got any better. Kept on through our stay at Yakima. Still really bad. At Fort Sill he still had them at it even though all the bugle calls were done by the Field Artillery Band.

Finally in Camp Howze, TX, after three years of trying in vain to teach his buglers he gave up. I was at Headquarters the day he blew up. One of these guys who had been going to classes on the bugle played "Boots & Saddles" call and Brownlow heard it. It was bad. He told Master Sgt. James Maloney (Cheyenne, Wyoming) to have the so-called buglers turn in their bugles to Supply. After three years he had had enough. Once in combat I heard him bitching about it. He said they didn't try and he couldn't do much about it. I always wondered if those guys got sent to the Infantry.

On some of the 10 minute breaks in the morning or afternoon, the sergeant in charge might say "Police up time!" Then he would add, "I don't want to see anything but assholes and elbows for the next 10 minutes. You pick everything that is foreign to the soil and put it in the garbage barrels."

Everyone went over the whole battery area and picked up cigarette butts, matches, paper, etc., even leaves. The way you get rid of a cigarette butt is to take the paper off and throw the tobacco in the wind, then roll the cigarette paper in to a little tiny ball and toss it. It will completely disappear. Cigarettes did not have filters in 1943.

We seldom had to "police the area" in other camps after basic training as prisoners from the guardhouse did it early in the war. Later, German and Italian prisoners of war did this.

There was an officer in Camp White who played bagpipes. Every night he was out behind the battery area playing the pipes. Well, he was not too good at it and everyone got to complaining about the noise he made. Finally Colonel Brownlow banned him from the battalion area and he had to go at least 500 yards out into the desert away from the battalion area to play.

And so he did. Every darn night you would still hear him pretty plain at our battery which was about ¾ of a mile away, especially when the wind blew from his direction.

There was a couple of old guys, one a sheepherder and the other a cowboy. They got along pretty good but did get into arguments about which was best, sheep or cows. They really got in to arguments about which towns back where they lived had the best girls and the best sporting houses.

Every few weeks there would be a full field inspection. All of the items of clothing, toilet, canteen and belt, shoulder harness, blankets, even rifle had to be laid out on the ground in a certain way the book called for (this book was the "Artillery Man's Handbook.") Everyone was equally spaced on the parade grounds for this even the non-commissioned officers. The officers would inspect to see if things were laid out right, and if each article was clean and polished.

These were the items you carried with you on a march. I never had a problem with this inspection but some did. They couldn't get things laid out right or the stuff might be dirty. Officers were really particular about the canteen, cup, mess kit, and the knife, fork and spoon. You had to keep them highly polished with steel wool. When you were out in the field, you could use sand from a creek to polish them.

When there was a barracks inspection you laid these things out on your cot. The cot had to be made a certain way and the foot lockers at the foot of the cot had to be arranged just so. Clothing had to be folded just a certain way. At the head of the cot was a shelf on the wall with a place under it to hang the clothes you were not wearing. These had to be hung all the same direction with the left sleeve to the outside, folded in to the front of the shirt or jacket so that all shoulder patches, chevrons, and hash marks showed. The extra pair of shoes had to sit in a certain place. "By the book, by the book."

I never had a problem with this inspection, either, but some did, especially hanging the clothes up so they looked neat and also keeping the foot locker looking neat. After basic training these inspections were held less often.

The weeks went by, practice, practice, doing things over and over every day.

At the end of thirteen weeks we had completed basic training. Privates were promoted to Private First Class, got a raise in pay and a stripe on the sleeve. We were now ready for our next assignment.

THE 242nd IS ORGANIZED

After basic training was over in December, the batteries were formed according to the Army TO (Table of Organization).

TABLE OF ORGANIZATION

ARMY – Usually 2 or more Corps

CORPS – 3 or more Divisions

DIVISIONS – At least 3 Regiments

REGIMENT –3 or 4 Battalions

Artillery Battalion – Five Batteries of about 100 men each

The Batteries were Headquarters and Headquarters Battery, A, B, and C Firing Batteries and Service Battery. The batteries contained the following: Headquarters & Headquarters - Kitchen Section, Machine Gun Section, Motor Pool Section, a small Medical Unit, and a Piper Cub observer plane.

The firing Batteries, A, B, and C, consisted of 4 gun (Howitzers) sections of 7 to 9 men each, a machine gun section of 10 or 12 men, a wire section, radio section, survey section, forward observers, battery agent or messenger, kitchen section, and a motor section.

The Service Battery had a kitchen section, motor pool, machine gun section and the Supply section for the 5 Batteries.

Headquarters & Headquarters Battery contained the Commanding Officer and Executive Officer or second in command usually a Major. He is probably a lawyer and is in charge of legal things. There is a Captain that is the intelligence officer, and a Captain that had control of communications and the Fire Directions Center. There are two or three lesser Lieutenants, also a Sergeant Major, the highest rank of non-commissioned officers (who if he is old regular Army practically runs the whole battalion of 500 men even though the commissioned officers don't realize it!).

We had such a Sergeant Major, Sgt. Maloney, Cheyenne, Wyoming, who later received a battlefield commission and became a 2nd Lieutenant. He had been in the Army 17 or 18 years, always Field Artillery, from horse drawn, mule pack to truck drawn and self propelled. He was a short pudgy red-faced Irishman who was sharp mouthed and always right. He knew everything about the Army and Artillery and military law, court martial, food, sanitation, anything that came up. Yet he was not as tough as he acted, there was always a twinkle in his eye even when he was chewing someone out even when he was telling a commissioned officer he was wrong, calling him "Sir" at the same time. He even set Major Brownlow right when he was wrong and Brownlow did not like it (although he may have appreciated it later.)

The second part of Headquarters and Headquarters Battery was the officers and men who did the radio communications and telephone lines between Headquarters and the actual firing batteries that had the 105mm Howitzers, A, B, C, and also to Service Battery. This part of Headquarters did survey, maps, overlays on maps, and contained the Fire Direction Center that directed and massed the fire of the twelve 105's. They had a full set of trucks and truck drivers, drivers for officers' cars and Jeeps, plus a section of 50 cal. machine gunners for protection of the Fire Direction Center and Communications Center and the big brass in general.

Plus there was a Message Center, which had 3 men from Headquarters, 1 each from A, B, and C batteries and 1 from Service Battery. Those from Headquarters were sergeants and the ones from A, B, C, and Service were corporals.

And finally Headquarters & Headquarters had a kitchen staff, a mess sergeant, two tech sergeants and 2 corporals. These five were cooks and they had PFCs and Privates assigned each day to help them. These were the KP, (Kitchen Police) and there were usually six of them. They did all the scrubbing, dishwashing, potato peeling, stirring the food on the stove, etc. The KPs did all the hard work. At least 2 cooks were on duty from about 4:30 AM to about 8:30 PM each day. Every private and private first class got their turn at KP, usually for 1 week every 3 months. Some did KP more often when they got extra duty for punishment such as being late for roll call, getting back late from a pass to town, not in bed at bed check time, sassing an officer, improper dress, bed not made up right, messy shoes, clothes or locker in a mess, etc.

The kitchen force and KP were duplicated in all 5 batteries.

Service Battery: This one is about 10 less men in strength. It has the job of supplying itself and the other four batteries with food, clothing, PX goods, guns, Jeeps, trucks, Howitzers, gasoline, extra transportation, anything the other four would need, even tools, spades, shovels, sand, gravel, lumber, anything. This is a tremendously heavy job, bad enough stateside but really hard, long hours in combat. A Captain, 2 or 3 Lieutenants, a first sergeant, four or five lesser sergeants, 8 or 10 corporals, and around 70 PFCs or privates and also a man called a Warrant Officer who is between an enlisted man and a commissioned officer head this battery. He is supposed to be an expert on supply. He only rates one salute a day. A commissioned officer is saluted every time you meet him.

Then there are the three firing batteries, A Abel, B Baker, C Charley. Each of these had about 100 men, including officers. Commanding was a Captain. A First Lieutenant is the Executive Officer who has charge of the men who fire the 105mm Howitzers. There were four of these gun sections in a battery and a fifth section that hauled the ammunition for the other four and also guarded the perimeter. This fifth section is also called the machine gun section. The Executive Officer has an assistant, a 2nd Lieutenant. There is one more 1st or 2nd Lieutenant that has charge of wire sections, radio section, survey section, forward observation, and the battery messenger or agent.

Another Lieutenant was the motor officer who was in charge of the battery trucks, Jeeps, trailers and their drivers, assistant drivers, mechanics and maintenance men.

Next in command was the First Sergeant who was much like the Master Sergeant at Headquarters. He actually ran the battery, roll call, passes, etc., usually an old regular army man who knew more than the officers and told them so, "Sir!"

Other Staff Sergeants were the head cook or Mess Sergeant, the Supply Sergeant who handled all supplies at the battery level. The Staff Sergeant who had the #1 Howitzer section was Chief of that section and being #1 was also called Chief of Artillery. Chief of the wire and radio section was a Staff Sergeant. Chief of the Survey and Fire Control and Message Center was a Staff Sergeant, also.

The 2nd, 3rd, 4th, and 5th sections within the battery were headed by buck sergeants. A Staff Tech Sergeant headed the motor section. Second cook was a Technical Sergeant.

All of these sections had corporals or technical corporals. Finally came privates first class, then privates in these sections. All people were trained to replace other people in their section.

In a Howitzer section, there was a sergeant, a corporal who was the gunner, a second gunner and 2 assistants. The two assistants and the other 3 men in the section were known as cannoneers.

Each driver was a Technical Corporal and had a second and a third assistant driver. There were always one, two, or even three backup people in case of casualties.

Also there was a technical sergeant who was an artillery mechanic or Howitzer mechanic, and he had an assistant that he trained.

That about covers the Table of Organization!

I was assigned to the Fifth Section of Battery A as a machine gunner. Several months later I was transferred to the Survey Section, made corporal and became the A Battery Agent.

CHAPTER FOUR

In January 1943, we got word we were to go to the IX Corps Artillery Firing Range at Yakima, Washington. Preparations were made that half of the battalion would go by truck convoy, staying one night at a camp in northern Oregon, Camp Adair, one night at a camp in Washington. The third day we would be in Yakima. The other half of the battalion would go by troop train. The train trip was to take 2 days, eating and sleeping on the train. I went on the train.

The troop train arrived in Yakima on January 15. We were taken by trucks out to the Firing Range, a tent city for about 2000 men. These were pyramidal tents with the famous Sibley stoves that burned wood or coal. Not a tree in sight—just sagebrush, desert, mountains, and snow. It was really a hole! The truck convoy arrived two days later.

The food was bad. It was too cold to eat in the mess tent. You would get your hot cereal, coffee, eggs and toast in the morning – by the time you ran back to your tent, the cereal would be froze and the coffee cold. You would need to reheat it on the Sibley stove in your tent.

It was 10 to 40 below zero every night, might warm up to 10 or 20 above in the daytime.

The food at noon and again at night was usually mutton stew that stunk so bad you couldn't eat it. Everyone lived on candy bars from the PX. I got up to 185 pounds at Yakima!

We went into town whenever we could get a pass, just to eat! Had to hire a taxi for the 10 or 12 mile trip! Cafés in town had good food. Most of us went into town on weekends and stayed all night at a hotel, except for the ones on guard duty.

The town was a pretty wild one, even the cafés, lots of fights. We had several Oregon lumber jacks in our outfit and they were all muscle. Gus Zetterberg, Silverton, Oregon, a Swede about 6 ft 6, 190 pounds was sitting in a booth eating one night in a café and some guy walked up and smarted off to him. Gus never even answered him or got up. He hit that guy while still sitting in the booth. The fella slid clear across the room on the floor and was out like a light.

Some of the lumberjacks liked to hand or arm rassle. Nobody but another lumberjack had a chance with them. One of them was a tree topper. His work had been to climb to the top of the high trees after the limbs had been trimmed off and cut off the top of the tree. In tree topping contests, he claimed he could toss his hat off from the top of the tree, and beat it to the ground with his climbing belt and boot hooks! We thought he was windy but the other woodsmen said it was true.

In Yakima we went to the bars, the cafés, the USO, the movies but mostly to the cafés to eat, and the hotels for a warm room and a good bed to sleep in.

One old sergeant named Martin Rauch (Wahnoka, Oklahoma) had all his teeth pulled in Camp White and had gotten false teeth. He got crocked one night, got sick and threw up his teeth in the toilet at the hotel and flushed them down the drain. He was without teeth again for about 6 weeks.

While we were in the States, once a month you had to fall out naked with just a raincoat on and line up by rank, alphabetically, and march to the dayroom to see the Battalion doctor. One by one everyone took off his raincoat in front of the doctor. He checked everyone for head lice, and crabs or crotch crickets as they were called, also for hemorrhoids, hernias, and venereal disease and athlete's foot fungus. In all the time in the States, I only remember a couple guys with head lice.

They shaved their heads and eyebrows clean and put some kind of ointment on their heads. Only one guy had crotch crickets and they shaved his bottom and used something called “Blue Vitrol” on him for about a week.

Most of this outfit were farm boys and pretty straight laced and did not fool around with streetwalkers or the houses of ill repute. The west coast was as bad or worse than the Deep South for this kind of thing. The few we had that got in trouble with venereal disease were mostly loggers, miners, sheepherders, and cowboys. This was a way of life they grew up with - - payday, booze, gambling and whorehouses.

At the showers on base there was a fair-sized tank or vat about 9 inches high going into the shower room that was filled with a chlorine solution. You had to step into this before going in to shower and again coming out. This kept athlete’s foot from spreading.

From the Firing Range at Yakima you could see Mount Rainier always snow capped off to the Northwest toward Seattle. Off to the Northeast was a sharp pointed ridge of mountains called the Sawtooth Mountains, completely covered with snow at this time of the year. Other mountains surrounded the sagebrush desert of the north part of the valley to the Yakima River. Near the river and south of it, most of the hops in the United States were raised. Hops grew on funny looking giant trellises. Also the world’s best apples were raised here.

It was always cold here that winter, below zero most of the time. At one time we had snow three and a half feet deep. When we went on hikes, we had to take turns breaking trail.

There was never enough coal to keep the Sibley stoves going in the tents. At least not as hot as everyone wanted. Each tent was rationed to so many buckets of coal a day. At first this was on the honor system but the coal disappeared faster than it was hauled in. Everyone was stealing it and hiding it under their bunks in the tents.

There were seven to eight men in a tent. The tent was pyramidal, about 12 x 12, over a two by four and wood wall about 5 feet high with a wooden floor over stringers of 2 x 6's. There was no way a little Sibley stove could keep this warm. You had to sit next to the stove, and the side away from it would get cold. The stove did not warm the bunks around the wall hardly at all.

Some guys had gotten their stoves so hot that the stove pipe set fire to the tent part and it would burn up in just a few minutes.

Old Joe Freas (Santa Clara, California) kept our fire going most of the time. About 2 hours after we had all gone to bed, it would go out. Old Joe got up at 4 AM and started it going and had water boiling to shave with. We also used the stove to reheat our coffee and cereal that froze coming from the mess tent to our tent.

They put a 24 hour guard on the coal pile and you had to have a ration ticket to give the guard to get a bucket of coal. All the tents were searched and the coal found under the bunks had to be returned to the pile.

So then the big deal began to be to get a guy on guard at the coal pile from your battery. Then he would let you steal coal while he was on guard at night. We pulled up some floorboards and packed extra coal under the floor. There were more searches under bunks but no officer ever thought about looking under the floor.

It was so cold sleeping in those tents that we put old papers and cardboard on our cots, laid our bed rolls on that and put 2 wool blankets inside the bed roll. We put our heavy wool overcoats on top of the bedroll, and a rain coat on top of that. Then you crawled into that with your 100% wool undershirt and under drawers and socks on. Still your side next to the outside wall would get pretty chilly.

When outside you had on all the clothes you could get on and if you kept moving you could keep warm except for fingers and toes.

In January it did not get light until about 8:30 AM and was dark again 4:30 or 5 PM.

We did much artillery practice, going into position, setting up communications, survey, etc., setting up machine gun posts, etc. A few of these deals were done at night. Machine guns were fired at towed targets at night. Also, some anti-aircraft guns were there that fired at towed targets at night, as well as in daylight. There was a battalion of searchlight people with the anti-aircraft people that lit up the towed target at night.

We were in Yakima about 6 or 8 weeks and spring was coming on. The weather was warming up and the snow was gradually disappearing.

We got word that we were to have men pulled out, officers and non-commissioned officers, for a cadre to start a new outfit somewhere in the South. The colonel chose all the officers and sergeants and corporals he did not like, or those that had been in trouble and promoted them to whatever rank was needed for the cadre and shipped them out. He also shipped out the bagpipe player.

Earlier we had some guys joyriding in a Jeep from our battery and they wrecked it, were broken in rank and court-martialed. Our Sergeant MacFarland was busted to a private. (Henry Holmes from Taylor, Nebraska was not hurt too bad; Albert Tobiason from Franklin, Nebraska had his leg banged up real bad.) In the cadre that went out, I lost Sergeant MacFarland because he had been involved in that Jeep wreck. They promoted him back to sergeant and shipped him out.

The Yakima River was a rushing mountain stream, clear water with a rock and smooth stones bottom. There was a shallow beach place where we went to wash the trucks and Jeeps. We just pulled out into the shallows and a bunch of men with rags waded in and washed the vehicles. Everyone was told, “Don't drive out too far because it gets deep real fast.”

So the first damn thing, Pansanante who was a loud mouth Italian, drives out too far in a big 6x6 truck, and it starts floating down the river, him in the driver's seat yelling his head off. After about a hundred yards the truck caught on a rock and hung there. They had to run a steel cable out to it and winch it back to shore.

There was a big stink about it for several days. And of course, when we sent out the cadre a few weeks later, they sent him on it and got rid of him, (like so many other screwups and people the higher ups didn't like.)

Some of the guys that came from Camp Roberts were promoted to replace those that left, some of them to “acting” corporal or sergeant (this meant you had the job but not the pay). This kind of set the scene for most of us that we would never get a very high rank because all ranks were filled for the duration of the war unless we should send another cadre out or someone got busted, was discharged or died. Well, we never did send out another cadre so the ones that did get a higher rank like myself got it through others being discharged or busted in rank.

We were told that we would be moving out also and we started loading our trucks, Jeeps and trailers, and cannon on flatbed railroad cars, blocked the wheels and chained them down.

Freezing in Yakima

Edward Walsh, Gresham, Nebraska

Ninth Corps Field Artillery Firing Range

Yakima, Washington

Edward Walsh, Gresham, Nebraska

50 caliber machine gunner

Joseph Freas, Santa Clara, California

Assistant

Ninth Corps Field Artillery Firing Range

Yakima, Washington

CHAPTER FIVE

We were to be packed and ready to move out at any minute, but several days of waiting went by before two troop trains arrived. We were hurried to the depot about 6 in the morning.

We stood and waited until 11 o'clock to get on the train. Same old Army deal, hurry up and wait. Half the men and half the equipment were to go on one train and the rest on the other. Finally got checked off by name, rank and serial number and got on the train. We pulled out of Yakima the 27th of February 1943.

We were all in Pullman cars, just as we were from Camp White to Yakima. Each had his own berth. There was a kitchen car to prepare our meals.

We were under sealed orders from 9th Corps Headquarters at Fort Lewis, Washington. The train I was on traveled across Washington to Twin Falls, Idaho. The other train went north and east along the Canadian border. We went across Montana and into Wyoming, then to Colorado, Kansas and finally Oklahoma. This took several days. On March 4, a cold wet morning, both trains pulled in at the same time at the depot at Lawton, Oklahoma, just a few miles from Fort Sill.

We were put on trucks and hauled to our new home, another tent city at the west end of Fort Sill's main camp. Our equipment came onto the base on a railroad siding. As soon as we were assigned tents and cots and were settled in, we ate and went to unload the equipment train.

Then there was a meeting. We were told why we were here and what we would be doing, the rules and regulations of the base, about guard duty, passes, where things were, the theaters, PX, cafés, etc. It sure sounded better than Yakima even if we were still in tents!

The reason we were here was to train the OCS, Officers' Candidate School (guys that came out of ROTC in college, or had been chosen out of ranks to try to become a 2nd Lieutenant in 90 days of tough training, or ninety day wonders as they were called). Fort Sill turned out officers from this school by the hundreds every month. Those who did not make it were given the rank of corporal and were transferred out.

Also at Sill was the Field Artillery School where guys were sent from outfits all over the nation to become specialists in radio, telephone, survey, mathematics, truck mechanics, gun mechanics, ordinance, ammunition, cooks, bakers and other jobs in the Artillery. We were to train these guys, too.

The Field Artillery School would furnish men to do jobs in their specialty for us and the OCS would furnish the students (would-be officers). We would do some work but mostly watch and help them or show what to do so it was right.

The next day we would go out and start "The Battle of Fort Sill", which would go on day and night for the next year.

The next morning we went on the firing range, not with our 105mm Howitzers but with French 75mm. Seems the Arsenal had millions of rounds of 75mm shells left over from World War I and they figured they might as well use up the ammunition.

It was cold and windy. They were short of telephone operations so I tried running one at the gun site. The whole thing was being run by telephone. I would take the order from the Officer Candidate student that was out front somewhere, as Forward Observer and I would repeat to the student at the French 75. Such as, left, so many degrees, right, so many degrees, up or down so many degrees. But I forgot to say "On the way" when they fired and the Forward Observers would not know the shell was coming and would not see exactly where it hit and could not give the order to change the degrees right or left or whatever. So after this happened a few times, I was off the telephone! I did run it lots of times later without any trouble, but I usually did recording of the guns.

We would fire from a few rounds up to 2000 rounds a day. I recorded day after day for A Battery until I was transferred to the Survey Section as messenger or agent.

The 75mm ammunition we had was bad. Sometimes it would not fire and the gun would have to set for some time to see that the ammo did not go off. Then it was unloaded very carefully and the misfired shell put off at a distance to be destroyed later. Sometimes the shell would explode just as it left the muzzle of the cannon. This is called a muzzle burst and is very dangerous. It could kill some of the gun crew but we never had anyone hurt.

I saw several of these muzzle bursts from the old French 75 but only one from a 105mm Howitzer with new ammunition. After a few weeks we finally ran out of 75mm shells and went to our regular 105mm.

The food in Fort Sill Tent City was much better than Yakima Tent City. About the only thing that was bad was greasy pork chops and also the coffee was bad. There were cafés on the post, though, and you could go into Lawton any night there wasn't night problems or guard duty. Some Chinese ran a café in the main PX. Some of us always went there for breakfast on Sunday and had steak and eggs and hash browns. You could buy most anything to eat in the PX and also beer in the evenings.

We had to have drivers' licenses for the Army vehicles and we had to have good control over what we were driving in order to get the license.

The Motor Officer would take a bunch of men and different kinds of trucks out to the back side of Signal Mountain. Here he would work with each guy in different vehicles and ride around. Finally in a place that seemed almost straight up, you had to stop the truck, set the hand brake, and shut off the engine. Then start the engine, put it in gear, release the handbrake and go on up the steep grade – without rolling back more than 18 inches.

Some guys never did get a license. The Motor Officer was a great big young guy just out of OCS. He was real easy to get along with, but he worried that someone would panic and that the big truck would roll backwards down the mountain, wrecking the truck and possibly injuring himself and the driver.

The Lieutenant sweat pretty easy anyway and it was hot the day I went out to get my license. His whole uniform was soaking wet. I was one of the last to drive and as I got in the truck, he said, "I'll sure be glad when this is over, it's wearing me out!"

We got to the grade and I stopped as directed. He said, "For God's sake, don't let it roll back or kill the engine."

I said, "I won't. I know how to do this". I put the truck in gear and pulled out. Never rolled back an inch. I could see the relief on his face.

The others said he had said the same thing to everyone on the grade, but some had rolled back 15 or 20 feet and scared him to death. He never went overseas with us. The Colonel must have thought he wasn't tough enough, or maybe too big to be a forward observer, would make too good a target for the Germans.

Lawton was a pretty tough town. The bars, cafés and even the theaters, bus and train depots were rough. This had been an Army town near an Army post for years and years. Back to the time of the great Indian chief Geronimo, and even before, there had been infantry and artillery at Fort Sill. There were signs on the lawns "Soldiers and dogs, keep off the grass!"

There were lots of Indians in the town and also lots of blacks on the "other side" of the tracks. In the 1940's the blacks still took the outside of the sidewalk for a white person. They also used the back of the bus and had separate rest rooms.

Some of the bar fights were really bad, knives, bottles broken over heads, broken tables, broken chairs and windows. Some guys got hurt pretty bad at different times.

I usually went to every change of movies on the post at the big theater. If I missed a movie there I could go to another theater on the base and see it later. I went to the show at least 3 times a week.

Fort Sill has a museum in the old original guardhouse. It tells the history of Fort Sill and has many old artillery pieces, etc. The cell in the basement where they kept Geronimo is available still to see. The legend is that Geronimo escaped from here once and rode a horse up on one of the bluffs of Medicine Creek. There are four bluffs right together, Medicine Bluff one, two, three and four. Artillery called them MB1, MB2, MB3, MB4. You can walk up these high bluffs, or ride a horse up as he did on the south side but when you get to the top it is a sheer drop on the north side of about 175 to 200 feet into Medicine Creek. Geronimo is said to have jumped his horse off into this creek and survived!

The solders chasing him were afraid to follow and Geronimo escaped into the Wichita Mountains to the west. However, he was captured again and lived out his life in the guardhouse or in and around Fort Sill. The military got so they didn't keep him in the cell and he wandered around the Fort, drunk whenever he could get some whiskey.

Most of the Creek that I saw was carved out of solid rock with deep holes in it. You could see the catfish because the creek was clear over the rock. It must have been that Geronimo knew where one of these deep holes was when he jumped off the bluff.

To the northwest was Mount Scott. To the west was Signal Mountain and beyond Scott and Signal were the rest of the Wichitas. These mountains only run for 15 to 20 miles west and only about 6 to 10 miles wide. Kind of a strange thing with flat plains planted to cotton, and pasture all around them. The mountains were covered with pine, scrub oak, brush, grass and rattlesnakes.

We lived in the tent city at the west end of the base right next to the firing range for about six weeks or more. We then moved to the big stone barracks where the artillery outfits that had been here for years had always lived.

They were crack outfits and were sent overseas. They even had a battalion of pack artillery that packed Howitzers on the backs of mules. They were called Mountain Artillery because they could go where no truck, tank, or even horse drawn artillery could go. This was a tough bunch. They went overseas to Burma.

Anyway, we got the fine stone barracks, steam heat, hot water, pretty plush. We called the stone barracks, "The Marble Mansions". We would be in these barracks almost a year. (Then we were back in the tarpaper shacks the last week or so in Fort Sill while waiting to go to the next post.)

These large stone buildings were Spanish style with lots of cross ventilation and large open porches or verandas on the north side and were real comfortable most of the year. There was about 2 months in the summer when the days were 110 degrees and 90 degrees at night with high humidity coming up from the Gulf of Mexico. It was bad enough when the wind blew to cool you, but when there was no breeze, you sweat just sitting still. You sweat so bad at night that the bed would just be soaked.

We thought this was bad but heard from guys in Arkansas and Louisiana swamp areas. That must have been worse because they had mosquitoes really bad. Mosquitoes did not bother us so much but we did have trouble with chiggers and ticks.

The food was even better in the stone barracks. Maybe they just had all the greasy pork chops used up. But we also got a new cook. In fact, 2 new ones, one a French chef, Joseph Monet (Cleveland, Ohio) and a 2nd cook who had been a chef, a Dutchman called (Victor) Fletcher (Jackson Heights, NY). They could put up a real good meal when they had the supplies. Like at Christmas, Thanksgiving and New Year's. They even had fancy menus printed. I have some of the menus in my suitcase yet.

We always had cold cuts on Saturday and Sunday nights because so many were gone on pass. Had a big dinner on Sunday noon though, of course, many of us got up too late for Sunday breakfast and went to the Chinese place.

Breakfast in the States usually was bacon, eggs, cereal and toast, real butter and orange marmalade. About once a week we had ham, eggs and maybe American fried potatoes, and once a week there would be the famous chipped beef on toast, and an apple, orange, or other fresh fruit.

When they had fried eggs, the eggs were practically raw! I could hardly stand them. When I did not take the kitchen job that was offered to me by Sergeant Reed, a friend of mine took it, Laverne Riel from Milford, Nebraska. When he saw me coming through the line, he would put two eggs over on the back of the grill and I would get some decent fried eggs!

(Riel married a girl from Medford Oregon that he met when we were in Camp White. Through attrition he finally became head cook and was a Staff Sergeant.)

The officers had a separate mess hall at Camp White and again here at Sill. They also slept in a separate building from the enlisted men. Base Command ran these mess halls and cleaned the officers' sleeping quarters. The Base Command furnished the sergeants and corporals and officers over them to do this. Most of the KP work was done by prisoners from the camp guardhouse and later prisoners of war, Italian and German.

Fort Sill had some huge mess halls that were fully staffed for the Officers Candidate School and for the Field Artillery School. There were normally about 80,000 men at Sill all the time.

One Sunday I was C.Q. (Charge of Quarters). This was a job that a corporal or any non-commissioned officer did about one or twice a month. What you did was stay in the First Sergeant's office for a 24 hour period and run errands, answer the phone and see that different work details and guard people got where they should be on time, take care of passes for guys to go to town, etc. Most everybody had a Class "A" pass by this time and could go anywhere anytime without permission.

They needed this pass to show to the post gate, or the M.P.'s in town that might be checking passes, but they had to sign out at the First Sergeant's office where they were going.

This office was called the Orderly Room or O.R. The C.Q. had to take his bedding and sleep there on a cot. The C.Q. also met any visitors that came and went, and found who they wanted to see.

So this Sunday evening, the phone rang. It was Post Headquarters and there was a forest fire on the west end of the reservation! They needed 8 men from every battery in camp. They were to be at Post Headquarters as soon as possible with a vehicle and a licensed driver. Well, there wasn't anybody in the barracks. I finally got seven, found another corporal who would take over my C.Q. job for the rest of the 24 hours and I went out to the fire.

There was truck after truck leaving Headquarters. We got to the fire in about an hour and a half. It was dark by now and the western sky was all red. Luckily this area did not have real tall pines. It did have quite a few small ones and lots of scrub oak and brush. Some of the first guys there with the direction of Forest Rangers had built a back fire, and it was backing up to a road and would be okay if the wind did not change, or if the fire did not jump the road.

Our job was to see that it did not get across the road. The trucks dropped off men with shovels about every 20 to 25 feet for three miles and they started spading the grass under along the road.

There were deer, coyotes, squirrels, skunks, bobcats, possum, coon and everything imaginable coming across that road, also rattlesnakes! The biggest one I saw was a big, old diamondback about 8 feet long, and was he mad, and in a hurry! We just let him go on across the road!

The wind did not change and everything was pretty well under control by morning. We were relieved by another truckload of men. We went back to the barracks and went to sleep.

The danger of the fire was not over for about a week. The area was kept surrounded by men on guard in shifts day and night all that time.

The barracks just west of Battery A was where the Field Artillery Band stayed. Many of these men had been in the Army for years and they were all professional musicians. They spent their days practicing music and marching. They did the flag ceremonies morning and evening. We all fell in to stand retreat as they lowered the flag at night. They did all the bugle calls during the day and they also played and marched in all parades and mass marching shows on Saturday afternoons. It was an honor for any new guy to get assigned to them.

Some of the parades were on the parade grounds north of the barracks and were quite a sight. The big ones were usually on Saturday afternoon when some big brass was visiting the post, like three or four star generals. There would be as many as five thousand men with colors, guidons (banners) and bands passing in review by the reviewing stand.

Some Saturday afternoons there would be mass artillery demonstrations for visiting brass and foreign brass. The impact area was near Signal Mountain. The people watching sat on the side of a mountain about a mile or a mile and a half away. There was usually over one hundred artillery pieces firing. Everything was surveyed in before hand and the shells would all hit in the same spot.

The cannon would be 105mm, 155mm, and 240mm. The minute they started firing on this spot, they would fire 8 to 12 shells from each gun, which would be around 1000 rounds hitting in the same place.

This was timed in coordination with medium size bombers with hundreds of anti–personnel bombs and then other planes with 500 lb bombs. The planes would get to the target just after the artillery had hit. They would drop their bombs in the dust and smoke caused by the artillery. Then fighter planes would come in and strafe the area. As the smoke and dust disappeared, just a few hundred yards from the impact area, infantry would come out of foxholes and advance across the area firing rifles and machines guns and bazookas.

This was really a surprise the first time I saw it because I did not know the infantry was out there. We were usually part of the artillery firing in these demonstrations, but one time we got to sit on the mountain and watch. It was really some sight. You could feel the heat from the bombs and shells and the wind pressure from it even a mile and a half away!

This had to be real accurate and precise. The infantry out there in the foxholes would have been wiped out if there had been a mistake.

CHAPTER SIX

A normal day at Fort Sill went something like this: Reveille at about 5:30 AM, fall outside and line up for roll call at 5:45. Breakfast at 6:00. Then everyone got what equipment or coat, etc., they would need for the day, and either go back to their bunk or go to the day room. They might sit around or lay down on the floor and go to sleep again, until first call at about 7:00 or 7:15. Then we fell out again, lined up and different groups were marched off to different details or classes. Maybe all would have half an hour of calisthenics or maybe an hour march or an hour of foot drill.

Now and then there would be a day of an hour of this and an hour of that but these were very few at Sill. Such days were usually a review of what we had in basic training and at Yakima Firing Center. Mostly we all went to our Sections and went out in the field. Each section took FAS students or OCS students and did everything that might be done in combat.

One of these drills was called "Reconnaissance, Survey, and Occupation of Position." We would get in our trucks, Jeeps, and command cars and leave the post by the main gate and head west on the highway along the side of the military reservation for a mile or so and then stop, well spaced in case of air attack. One or two Jeeps would go ahead several miles, enter the reservation and look around to find a make believe enemy. They would find a location to put our cannon and a place for the perimeter machine guns for defense of the position. They would radio back to the group or send someone back if we were under radio silence.

Another Jeep or weapons carrier with the Survey Section would go up there and put in survey stakes where the cannon should be for all 4 guns of each battery, A, B, C. They would figure out the angles of elevation, wind, humidity, etc. so that the guns would hit the enemy the first time they fired. All 12 guns would hit in the same spot the first time. The enemy would not know this "fire for effect" was coming, about 3 or 4 rounds from each gun or 36 to 48 rounds.

In the meantime the Communications Section has come up and run telephone wires to each firing battery and back to a protected spot where a fire direction center was set up, also a headquarters. The wire would be run to the forward observer who is located where he can see the enemy, at the top of a hill, or some such spot, or maybe a gully out of sight of the enemy.

Howitzers fire up and over a hill like throwing a ball over a house, kind of a lob. They do not fire in a straight line like a rifle or machine gun.

Meantime the bulk of the battalion is still parked along the road back several miles. Most are sleeping with a few guarding and a few watching for hostile aircraft with field glasses.

Now after two, or three hours the survey is complete. Communications are nearly complete and the forward observers from each firing battery go to their observation posts. They each have a Jeep to go part of the way. The Jeep has a mounted radio and the driver operates the radio and stays with the Jeep. The forward observer and a sergeant or corporal crawl up to the observation point with field glasses and a walkie talkie which can be used until the telephone wire is complete back to the Fire Direction Center.

A bunch of men from each battery is then hauled up and set up as guides at crossroads and so forth where the guns are to go. Then come the firing batteries and they go into position and "lay in" the guns or gun sights according to the Survey.

All four cannon are "lined up" on the aiming point. At Camp White there was a small mesa or plateau just across the Rogue River with one pine tree (Lone Tree on Tablerock) and this was used as an aiming point. At Sill there was a blockhouse on Signal Mountain, a shack on Medicine Bluff 4, and other spots. This doesn't mean the cannon were aimed at the tree. It means the gun sight, which is an optical periscope. If there is no point available such as a flagpole, the corner of a building, anything that is straight up and down, one of the cannoneers runs out about a hundred yards in front of the cannon and sets up a tall red and white stake, called an aiming stake.

The gun sights on each of the four cannon are on the aiming point. About 25 yards behind the cannon between No.2 and No.3 cannons, the Executive Officer has an instrument called an aiming circle, which takes into account "true north", "magnetic north, Y Azimuth distance between the two" – the reading from this to the cannon and the aiming point. The officer takes a reading from this and gives a reading to each gunner that has his cannon at the zero deflection and a certain elevation from level.

The cannon were originally parked about 50 yards apart pointing in a compass reading off a map in the general direction of the target. When the one and two gunners set the deflection right or left and the elevation up or down, the four cannons are at the same elevation and deflection and are parallel with each other. This is called "base deflection" and "base elevation".

After every firing session, the cannon are returned to these readings and always start from here. Like, base deflection left or right so many degrees, base elevation up or down so many degrees. When they are all fired, the rounds will land at the same distance and 50 yards apart. The sights stay on the aiming point but the gun is traversed, or raised or lowered to the degrees called for from the forward observer, to the radio or telephone operator to the Executive Officer who yells them to the first and second gunner at the cannon. One gunner controls elevation and the other deflection and aiming stake.

The guns are dug in, foxholes are dug, and the shells are laid out to fire.

Meantime the rest of Headquarters Battery has come in about a mile or so back and dug in. All batteries have put out guards and machine gun nests and camouflaged the guns and trucks under trees or nets. Service Battery has pulled in a mile or so behind Headquarters and set up guards and camouflage.

Service Battery has all the extra supplies not carried by the firing batteries and Headquarters such as food, shells, gasoline, parts, radios, telephones, etc. Each truck that pulls a Howitzer has 40 or 50 rounds for its Howitzer, plus the 5th Section, the ammunition section, has several truckloads of shells.

Finally everything is ready and when the Forward Observers think there is a bunch of make believe enemy in the open, the guns all start "firing for effect". When whatever is left of the enemy has taken cover, the forward observers pick a few targets, like "Enemy tank in the open, left 16 degrees, up 8 degrees, center, fire center guns once."

Telephone operators tell the forward observer "On the way". Shells hit.

Observer says, "Left 3 more degrees, up 2 more degrees."

Telephone operator says, "On the way." Shells hit and are close to target, or maybe they hit one on each side. Observer says "Bracket" or "On target." Then he might say, "Fire 3 rounds for effect." All 12 guns fire 3 shells and if the tank is knocked out, the observer says, "Cease firing. Mission accomplished."

In some cases you might have had each of the three batteries firing on separate targets because all three had forward observers and could work directly by radio, or through telephone switchboard with their own battery. Most of the time, though, it was through the battalion switchboard and all 12 guns were controlled by the battalion fire direction center.

The shells used on the "enemy" would probably have time fire, and the timers on the tip of the projectile would have been set to go off about 30 or 40 feet in the air above the enemy. If the ground were really frozen or rocky, a delayed fuse would be used. It would hit the ground, bounce, and then go off. A delayed fuse was also used on buildings, or on dug in enemy for more penetration. A "quick" fuse was also used to go off on contact like on a tank or truck, or on treetops above infantry.

All the above shells would be "HE" high explosive. The shells for tanks or pillbox of a fort would be "AP" armor piercing or "WP" white phosphorous, that sets fires or burns. WP could be used on buildings to set them on fire.

In addition to the lob, or loop angle, these Howitzers also fired a high angle, like almost straight up and a little forward and down. Like on a mountainside, you could fire almost straight up and the shell would come down on the reverse slope of the mountain. It was possible for the cannon to be fired directly at a target by turning everything over to the gunner on each gun. He had a sight that he could use to fire straight at something, like a tank or infantry, up to a few thousand yards.

I never saw that happen in combat but we came close to it on several occasions. Luckily the real enemy was stopped before they broke through that far. At other times we had time to retreat before they got on us.

This RSOP drill (Reconnaissance, Survey and Occupation of Position) might turn into a different situation. Perhaps the "enemy" was about to break through our infantry, they had little or no artillery and we are back a few miles in reserve. In that case, a call comes on radio to help in a certain area and the coordinates on the map are sent in code where to go. So the reconnaissance team and forward observers take off for the area fast. In just a minute or two, the three firing batteries take off, as fast as the trucks will go. The recon guys from each battery find a position for the four guns of their battery and guide them into position. "Occupation of position."

The forward observers for each battery have gone up to the infantry in trouble and call back a compass reading to point the guns in general direction and give an elevation and say, "Number one gun fire one round." The radio operator says "On the way", and the forward observer watches where it hits and calls back, right or left, up or down so many degrees, then fire one or maybe center battery one round. When he gets adjusted on target he calls for battery 3 or 4 rounds for effect on target. Perhaps more rounds are needed to eliminate target. Each battery and observer are working independently. This all takes place in a matter of minutes.

In the meantime wire sections are running wire for telephones to observers and batteries and Headquarters so all 3 fire batteries can be controlled by Fire Direction Center at Headquarters and more heavily concentrated shelling can be done.

In case the batteries would have to retreat, the call came out, "March Order, Abel, Baker, Charley", or some combination depending on who was closest to the enemy and was going to be overrun. This one would retreat first. This could be done in one or two minutes, ammunition loaded up, gun hooked to truck and moving to the rear as fast as the truck would go. In combat some things may get left behind.

Out in the field on the almost daily RSOP, the noon meal was prepared by the cooks out of the mobile kitchens and the men used their own mess kits and canteen cups. Leftovers went into a sump hole that had been dug in the ground. Two portable stoves had 30 gallons of hot water on them to wash the mess kit and silver, one barrel soapy, one boiling hot to rinse.

Straddle trenches were dug for going to the toilet. If the trench was too far away, you dug your own and covered it.

After the field problems were completed, say about 3 PM, we would be back in camp and all the equipment would have to be cleaned, and the vehicles washed. The transits, instruments and binoculars were dusted, wiped down and put in cases. The rifles were cleaned and oiled.

The gun crews would have to clean and oil the Howitzers. The barrels on the Howitzers were swabbed with a brush and soda water, swabbed dry, and then swabbed with an oily rag. This was done with a ramrod much like the old Civil War type, except on a 105, the rear of the barrel or breech called a sliding breech block came out, and you swabbed all the way through. One of the cannoneers disassembled the breechblock, cleaned and oiled it.

The machine guns were done the same way. The maintenance on everything was done this same way in combat, especially after any firing or road traveling. The drivers of vehicles carried extra oil, grease and gear lube with them. Everything was kept in top shape.

This was like the old Army Cavalry and Artillery, they took care of the horses first and then themselves.

When the Sections were done with their work, they were marched back to the barracks and dismissed. Everyone went to the showers and most everyone had to shave again. We then put on dress uniforms.

At 5 or 5:30 everyone fell out of the barracks, lined up and marched over to the flagpole and stood at attention through the nightly ceremony of the bugler playing retreat while the guard mount of the day lowered the flag and folded it.

Everyone was then dismissed and went to eat supper, or to the PX, or went to town on pass, or ate in the mess hall and headed for the picture show. Some might go to the barber.

There was a barbershop in our barracks in Fort Sill. Sometimes when I got a haircut I got a shave and a facial massage. The whole works cost very little. The barbers did not pay any rent and probably were subsidized by the Army.

After supper some of the guys rested on their bed and read books. A few were avid readers and were reading every minute of their spare time. It was easy to carry a book in your musette bag or the large pockets of the fatigue uniform. Some played pool in the day room, and some played cards, usually poker. A few days after payday the poker ended and they played pitch, rummy or bridge. Some were so tired they just went to bed and slept from about 7 in the evening until 5:30 the next morning. Some just sat around on someone's bed and visited, or told each other lies or shot the bull.

There were always long lines waiting to get into the theaters in the evenings, especially the nights that there was a new movie. There was great camaraderie in drinking beer at the PX (3.2 beer). Great friendships were formed there.

You could buy almost anything at the PX, beer, candy,

postcards, stationery, and toilet goods such as razor blades, aftershave, cigarettes, soap and so forth. The Army furnished laundry soap to do your own laundry, but no hand soap or toilet goods.

There was a camp laundry where we sent most of our clothes once a week to be washed. The laundry work was done by prisoners from the guardhouse who were guarded by MPs.

At nine o'clock, the bugler played lights out and the Charge of Quarters turned out the lights everywhere except toilets, hallways, day room and the First Sergeant's office where the Charge of Quarters stayed. At eleven o'clock the bugler played taps and everyone was supposed to be in bed. The CQ would go around with a flashlight and check every bed to see if everyone was there. If not, they were put on report and had to visit the battery commander next morning to explain why. There was usually not much of a problem with this.

In the basement of one of the big stone buildings was a very large room that had bleacher type seating in one end and the other end had a very large framework built up off the floor about six feet like a table top. It covered the entire end of the room. On this was a cover of netting or burlap and papier-mâché. It was a complete scale model of the Fort Sill Military Reservation, every tree, mountain, creek, and building. This was a training model for the forward observers.

The instructor would pick out a target on the model and have the student zero in on it from an aiming point or base deflection and base elevation. Under this table it was all set up in degrees of deflection and elevation. A couple of men were under it with a gadget that would put up a puff of smoke through the netting on the spot that the student called for. The gadget would put up one puff for one cannon shell burst, or two for center, or four for full battery or 12 battalion fire for effect.

It was just like firing out on the real military reservation and saved a lot of ammunition. This was in constant use in the daytime. It was called the "Puff Range". Some of us would go over and practice with it in the evening. We got real good at it. This helped a lot of enlisted men when an officer was hurt or killed in combat and someone had to take over for him.

One of the men went AWOL for a week or so and the MP's picked him up in his hometown somewhere in the South. He was brought back to the Guardhouse at Sill and was to be court-martialed.

When the day came, First Sergeant Reed called me into his office and told me to put on a dress uniform and go to the supply room and sign out a rifle and live ammunition. I was to go to the motor pool and get a driver and Jeep. He gave me the papers to do this and papers to "Pickup". This meant to get the man from the camp guardhouse. He was to be brought in handcuffs to a building on base at a certain time for the trial. He said I should be sure to do all the military courtesy saluting and so forth.

In the Jeep the prisoner sat in front with the driver and I sat in the back. Inside the court-martial room there were about eight or ten officers, three or four clerks, a sergeant and some corporals. I stood back of the prisoner's chair at parade rest.

One officer was his defense attorney, another the prosecuting attorney, one acted as judge, and the rest were the jury. There were arguments heard on both sides and the officers decided. Of course this was all done "by the book." There was a book for everything. (I never got to read the one on how to go to the bathroom but I bet there was one!)

The defense claimed his family was having problems at home and he couldn't get a furlough so he just went home on his own. Of course, you can't do this in the Army and the members of the court-martial ruled this out and found him guilty of being absent without leave (AWOL). The prisoner was to be kept in camp guardhouse for three months and fined two-thirds of his pay for these three months. It was over. I saluted the head officer of the court-martial and took the prisoner back to the guardhouse.

I turned in the Jeep and driver and went back to Sgt. Reed's office and reported in. He said, "How did it go, Kid?" I said, "I believe, perfect." He said, "Fine, Kid. I'm going to give you the rest of the day off. Just stay in the barracks or day room until after retreat at 5:00." The court-martial had lasted a little over 3 hours, so I had over a half a day off!

I never did know what news different ones got in letters from home, but I did see many sitting on their cots reading letters and crying. I don't know if girl friends had dropped them, or if there was sickness or death at home, or what. I did know of two whose parents thought that their sons were as good as dead when they joined the Service and this may have been reflected in the letters. Some may have been just plain homesick.

I really did not have these problems even though we had really serious problems at home.

Many weekends some of us would take a bus into Lawton and go to other towns, Oklahoma City, Annadarko, and El Reno. Salvador Talavera (Los Angeles, California) and I were in Oklahoma City one Saturday night and met up with Captain Robert Carlson (Minnesota) and his wife. Talavera was the captain's driver at camp. You are not supposed to fraternize with officers but he said, "Come up to our room at the hotel and we'll have a drink." So we did, talked awhile and when we were leaving, he said, "Here's a fifth of whiskey for you to take along. I had the bellhop get me three bottles from a bootlegger. I don't need it all."

Oklahoma was dry at that time. You could buy beer with 3.2 alcohol content but that was all. There was no hard liquor available in stores but bootleggers were getting rich!

The Captain was a great guy!

Edward Walsh, Gresham, Nebraska

Salvadore Talavera, Los Angeles, California

Fort Sill, Oklahoma

I went to Lutheran Walther League in Lawton a couple of times with Henry Duer from Staplehurst (NE). (A young people's church group) I didn't care for it and did not go anymore.

I had a boil on my wrist that I tried to heal with hot packs but it got worse and would not open up so it could heal. Infection set in, red streaks went up my arm so I went on sick call and the Battalion doctor said I had better go to the hospital. I was taken there and the doctors put something on the area that seemed to freeze it a little. Then they cut across it one way and then the other. Out popped a big green core about half an inch long and as big around as a pencil. They put some stuff on it, wrapped it up and sent me to a ward to have hot packs put on it.

An orderly went with me and he said, "Can you walk?" I said, "I think so", but by the time we got down the long hall to the ward I was about to pass out. We got in the door of the ward and there was this tall good-looking red haired nurse, a 2nd Lieutenant, and she said, "What the hell's the matter with you? You look like you've seen a ghost out in the hall." She set me in a wheel chair and had me put my head between my knees. By then I was almost out, but she started putting cold cloths on my face and the back of my neck and I got a little better. She and the orderly pushed me over to a bed and took off my clothes and put me to bed. I don't remember much after this. I went to sleep and was out of my head with fever for several hours.

Next morning I felt better and a different nurse said, "Get out of bed, go shower and shave, you look like hell." Later in the day, the red head was on duty again and she said, "You look better today. I think maybe you'll live after all." I said, "I don't think that is very funny." She said, "We will probably keep you here six or seven days until the infection is cleared up and the wrist heals."

I may have been there eight days. She was a pretty good gal but she sure bossed people around, the nurses under her and the patients. The patients that were out of bed had to keep the ward clean. Dust, sweep, and clean toilets and showers, and also the brass doorknobs, hinges, and the brass shelves and tables by the beds. They really shone. I don't know why they were brass.

I met this nurse on the street one time and saluted. She saluted back and stopped and asked how I was doing and if I had any more boils. I said no, we saluted again and went on our way.

That wrist was pretty bad but at least I didn't have boils on my rear like one of our truck drivers did. He had four of them and they lanced all four at once. He said he never sat down for over two weeks!

After being in the Service for one year, I finally got leave or furlough in August of 1943. Two weeks and travel time which was 3 or 4 days. I took a bus from Lawton, OK to Lincoln, NE. My brother Bob and my Dad picked me up. We came into Seward, and then Mom and Dad and I went out to the farm.

I did quite a bit of running around those two weeks, mostly with Puzz Menze (Gresham, NE) and his girlfriend, Doris Armagost (Surprise, NE). Also with Dub (Earl) Eppenbaugh (Gresham, NE) and his girlfriend, Jenny Weaver (Lincoln, NE). Went around to all the towns we used to go to before I left, Utica, Staplehurst, Surprise, and Shelby. I also went to some celebration in Dwight. But many of the people I knew were gone, guys to the service and girls to jobs somewhere. I did see Rosie Pokorney (Shelby, NE) in Shelby whom I had known for years. She always wrote to me while I was in the service.

I was too bashful to ask girls for a date. I had gone with Doris Schrader from Surprise (NE) a few times before the war and with Mary Sievert from Shelby (NE) several times. The girls I knew, and had gone to school with, that I might have asked, had moved away. Due to the Depression and drought, their parents had gone broke and had to leave their farms. They had moved to places where they could get a job to make enough money to feed their families. I did see a girl that went to the Ritchie School (rural Gresham, NE) who was in the 4th grade when I was in the 10th. She came up to me and said, "Do you remember me----------?" I said, "Sure, I remember you, you were just a little girl when you moved away from our neighborhood." She said, "Well, I'm a big girl, now!" I didn't take the hint.

I think I knew when I was on this furlough that World War II had changed the world forever. Nowadays you hear about this on TV and read it in magazines and papers, so when did they figure this out? I knew it in 1943.

Practically everyone you knew, even relatives had scattered coast to coast. Some I would see again, and many I would not. You always thought you would be a farmer. This idea along with the small farm was already beginning to disappear. So why worry about it? Maybe it's for the best.

When the furlough was over, the folks and I went down to Opal and Harry Pratt's in Friend (NE) for the night. Opal (Modlin) Pratt was my big "sister". She lived with my folks after the loss of her parents. The next day I got on the bus at Friend to go back to Sill.

Harry had a friend make a combat knife out of a large file and he gave it to me. I carried it all through combat and still have it. No, I never had occasion to use it.

The bus headed south down Highway 81 to Hebron, and on into Kansas. At Concordia, Kansas, a girl got on the bus and she could have sat anywhere on the bus. It was not that crowded, but she sat down in the seat next to me. In just a few minutes she had introduced herself and said where she was from. She was going to Wichita where she worked. We talked all the way to Wichita, about three hours as the bus stopped in every little town. We exchanged addresses and wrote back and forth for about a year and a half. She always wanted me to come to Wichita on a 3 day pass to see her, but I never did. Eventually she became engaged and the correspondence ended. We were like the two ships that passed in the night - but good pen pals.

Next day, I got to Lawton and on out to the Fort to do some more of the "Battle of Fort Sill."

Edward Walsh, Fort Sill, Oklahoma, Training

Albert Tobiason from Franklin County, NE, came back to us while we were at Sill. He's the one who was seriously injured in that Jeep accident at Yakima. He had a leg mashed and the hospital never got it put together right and he came out with a crooked leg, a limp and was also in pain most of the time. After being in the hospital for several months at Fort Lewis, Washington, he was sent back to the 242nd. They broke Tobiason from a PFC to a private and eventually gave him an honorable medical discharge.

After a few months, the other guys in the Jeep accident thought it was all pretty funny. All they got out of the court martial was broken in rank to private.

We got in several guys as replacements that had not made it in OCS. When they got booted out of OCS, they were sent to some regular line outfit as replacements for ones sent home as over age or medical discharges. One had an IQ of 160 so don't know why he didn't make it. Another hit one of the training officers and knocked him out for about 10 minutes. Another was just wild and crazy. He was with us for just a little while when there was a call for volunteers for Airborne troops and he went to the 101st Airborne. One of our guys got a letter from him later and he had gotten his leg broke on D Day in France. I don't know if he parachuted in or went in on a glider. In the letter he said he was going back to duty and hoped to get back into combat before the war was over. (We fired for the 101st Airborne in Europe and we kept an eye out for him and asked about him but no luck.)

Sometimes out on RSOP and other exercises we would sleep out in the field. Or maybe the whole deal would be at night. Mosquitoes were not really that bad but chiggers were, and also ticks. Also, you didn't want to put your bedroll down next to a rattlesnake. In that area they were mostly diamond back and big and deadly. Then, too, you had to find a place to sleep where you wouldn't get run over by the vehicles coming and going. A great big boulder was a good place, hard, but safer, and the chiggers weren't as bad as in the grass.

Many hours were spent waiting, especially after I was made agent in the Survey Section. We would get our surveying done and the rest was up to the gun sections, wire and radio sections and forward observers.

Hurry Up and Wait

Arnold Johnson, Franklin, Nebraska

Fort Sill, Oklahoma

Sometimes I would go visit with Arnold Johnson (Franklin, Nebraska) for a while. He would be running the battery switchboard that was always used here in the States. (He never got to set it up and use it in combat. Most of the time there was no time to run wire. If they did, it was a direct line to the guns.)

Arnold was kind of a sly comic. A lot of the guys had watches they were really proud of. Charles Dunbar (Alliance, NE) had a railroad watch. He had worked for the railroad and NO watch was better than his. It was a pocket watch like railroad workers carried. He claimed trains ran by this time because it was so accurate. Arnold would walk up to Dunbar and say, "What time is it by your watch and chain?" Dunbar would pull out his watch and tell him the time. Arnold would say, "Are the trains on time with the watch today?" Everyone thought it was funny except Dunbar.

There was a hillbilly kid that had bought a wrist watch (that was supposed to be THE thing at that time), and Arnold was always on him about his watch and chain, and if he checked anywhere to see if it kept time right. There were several others that thought they had the best watches but Arnold's old wrist watch was probably as good as any of them.

So my old watch gave out and wasn't worth repairing. I decided to get a real good wristwatch, anti-magnetic, waterproof, radium dial, and all that stuff. Henry Duer, (Staplehurst, NE) and I went into Lawton to a jewelry store and I found one I wanted. It was $70.00, which was a lot of money then, but I bought it anyway. It was okay for about 3 years. After the war, it quit. I took it to be repaired and they said it was not worth fixing, that I should not have paid over $35.00 for it in the first place. I knew that you should not buy anything near a military base. They are all out to take the soldier, especially in a town like Lawton. But it worked long enough to get through the war so I guess it was all right. After I got the watch, I started getting the question, "What time is it by your watch and chain?"

Christmas Greetings from Fort Sill, Oklahoma
From Edward Walsh to the folks back home, 1943

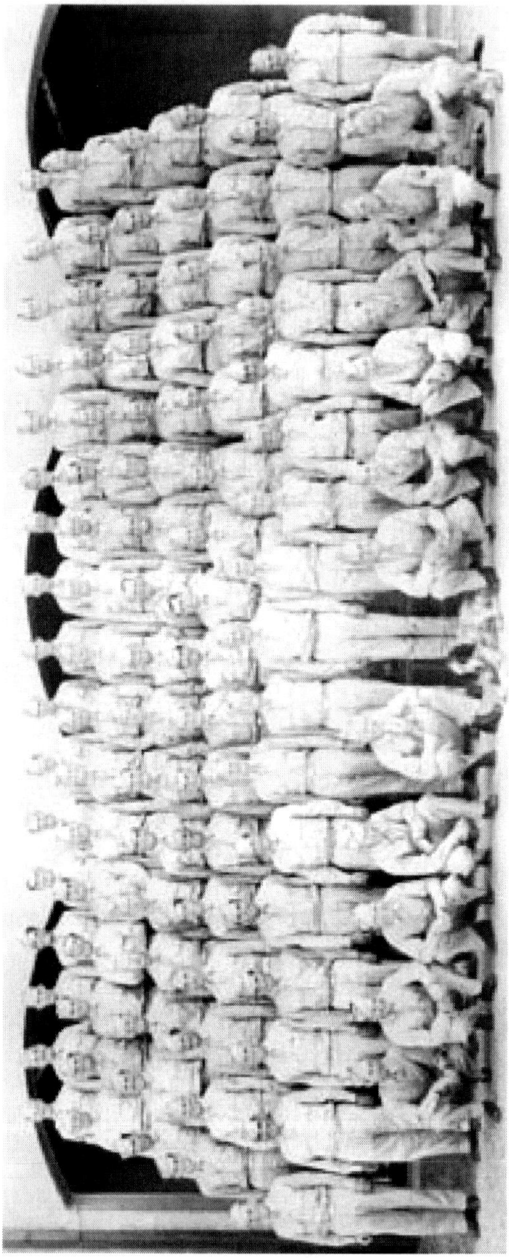

A Battery Group Photo – Fort Sill, Oklahoma

Edward Walsh, Gresham, NE top row far right

CHAPTER SEVEN

Toward the end of our time at Fort Sill, I walked guard for the second time ever. (I had walked guard once at Camp White.) This was at a time when we were getting closer to going overseas and the brass wanted to see what you would do in a tight spot. This time I carried a loaded rifle.

I was walking a guard post around some Howitzers parked in an old stable area of the old horse drawn artillery. I noticed some movement in the shadows of the stables. I stepped back in the shadows of the stable so I could not be seen and yelled, "Halt! Who goes there?"

I did not get an answer. I pulled back the bolt on the old Springfield rifle and threw one in the chamber.

"Halt, who goes there," I said again, and a man stepped out and said, "Lt. such and such, Officer of the Day."

I said, "There's two of you, the other one better answer." Someone said, "Col. Brownlow, Commanding Officer of the 242 Field Artillery Battalion."

I said, "Give the password." They did and I gave the countersign and said "Advance one and be recognized." He advanced. The second one advanced and both saluted.

They said I did the best job of anyone on the 6 guard posts that night. That was best out of 18 men. I believe they thought I would have shot them.

Once Walt Felhafer from Staplehurst (NE) and I walked across the base to see Alysius (known as Beans) Reinmiller, also from Staplehurst. He was an older guy who was a cook in the base kitchen. I believe he may have stayed at Sill until the war ended.

About a month before we left Sill we moved out of the Marble Mansions into tar paper shacks down in the OCS Training Center. The shacks held about 8 men with cots. No more beds!

Our equipment was moved out and lined up on a large oil mat parking lot. We had been storing it in the old horse stables and pens that the horse drawn artillery had used when they lived in the Marble Mansions.

We continued to train the OCS and FAS students but we knew we would be leaving Fort Sill soon. There were rumors of going to West Point Academy, or the Army Proving Grounds in Maryland, or going overseas. We had not yet taken the Army Ground Forces tests so it seemed unlikely that it would be overseas.

We got word we would move out of Sill by motor convoy. There was not much preparation for this. We just threw our bags in the trucks one morning and took off. I don't know if sealed orders were opened on the way, or if the officers knew where we were headed. No one in the trucks close to me knew. About 150 miles later we reached our destination.

The convoy pulled into Camp Howze near Gainsville, Texas at 1400 hours March 11, 1944. We were assigned to the 244th Field Artillery Group.

Howze (pronounced Howzie) was like basic all over again but with greater precision. Everything was honed down to a sharp edge to prepare the men for the Army Ground Forces Tests to see if they were ready, and good enough for combat.

There were many foot drills. Everyone, even the privates took turns calling the commands and counting the cadence. It was a good thing there were no cliffs around the camp. We might have been marched right off them. There were a few times when we didn't get the right command quick enough and got marched right up against a building. But everything smoothed out okay.

I went to First Aid School for a week and then taught first aid one hour a day for the rest of the time we were at Howze.

I also went to Chemical Warfare School for a week. You learned what to do in case you were attacked by chemicals especially gas, and how to protect yourself. We also learned how to wear a gas mask, how to test for gas, and how to clear a gas mask. We actually used the mask in a room full of tear gas. Better have it on right, or your eyes really got in a mess! The gases all have different odors.

We learned how to clean up a contaminated area wearing a mask and impervious clothing. The clothing was a coverall suit, hood and all made out of the rubber raincoat material that was used in those days. You had to spade under any places poison gas had splattered from a shell to stop the fumes and to keep from getting the gas on your skin. These gases caused terrible blisters in seconds on exposed skin and would kill in seconds if inhaled.

It was very hot doing those simulations in the Texas sun in those clothes and with a gas mask on. It was even hotter and more uncomfortable in what was called impregnated clothing. This was a regular fatigue work uniform that was impregnated with chemicals to make it air tight. It was stiff and really strange feeling and even hotter than the rubber coverall.

I also taught chemicals one hour a day the rest of the time we were stationed in Texas.

Everyone that taught classes had problems with ninety some men in a hot day room. Some would go to sleep. Some had smart remarks. There was always a Staff Sergeant to keep them awake or tell them to shut up and pay attention. All the other classes like machine gun, rifles, switchboard, telephone, radio, survey, slide rule, and so forth were taught by corporals or sergeants.

We had shots and inoculations here again. Seemed like it was about every three months. The entire battery was lined up the same as for pay day, by rank, alphabetically, single file and the medics gave us the shots. A couple times I almost passed out from these. Usually got one in each arm and maybe a third one. Many of us did not eat supper after having these in the afternoon, especially if you had a tetanus shot. In just a couple of hours you had a fever, sweat, chills that lasted all night. Your arm was so sore you could not lie on that side.

I had small pox vaccine three times in the Army and once when I was a kid and it never did "take". But the vaccination usually made me almost pass out. The ones for typhoid were bad also.

Some of our medics never did get very good at giving shots but we had one big fat medic that was really good. He would grab your arm, practically lift you off the ground, and you would hardly feel the shot. There was an old Mexican medic that was pretty good with shots, too.

Here in Howze the Officers' and First Sergeant's offices were in a tarpaper-covered building that also had the supply room, day room, toilets and showers. The barracks were off to the side and were long tarpaper covered buildings that held about 25 to 30 men. You had to leave the barracks and go to another building to take a shower or go to the bathroom. When I was "Charge of Quarters" at Howze I spent a lot of time trying to learn to type on the battery clerk's typewriter.

Right next to this building was a small PX with a large outside beer garden. Everyone that had any money was over there in the evening. I believe it was only open from 6 to 9 in the evening but everyone had a great time there. Staff Sergeant William Powell, (Sheridan Wyoming) Talavera and I went either there or into Gainesville to eat at the "Dumb Dora" café. The café had little square tables and red and white checkered tablecloths, good food and not expensive.

Harold "Dusty" Menze (Gresham, NE) was in the camp in a railroad battalion. Henry Duer and I visited him a couple of times.

Many of the men went on weekend pass to Denton, Dallas and Fort Worth. I never went any farther than Gainesville and did not stay overnight.

During one of the AGF tests, there was an accident involving one of our trucks. This was a 4x4 weapons carrier, a low, wide box pickup with 4 wheel drive. It was loaded with large spools of telephone wire, phones, switchboards, shovels and other tools. It usually carried 5 or 6 passengers.

The AGF tests were for speed and accuracy and we were going into a firing position as fast as the governors on the trucks would allow, probably about 45 miles an hour on open country trail, not even a trail really, across gullies and through brush. The wire truck made a turn too quick on a steep hillside. It rolled over, throwing men, spools of wire and equipment everywhere. One of the men was killed and two or three were injured.

Sgt. William Powell (Sheridan, Wyoming) was the head of the Wire & Survey Section and as such according to regulation accompanied the body home and had to stay for services. When he got back, he said it was the worst thing he ever had to do. The soldier was an Arab, and the strange religious service and the middle-east type of wailing had really been difficult for Powell.

In March I came home on emergency leave because my mother was very sick. I believe it was for 14 days and travel time. The Red Cross people put me on the train and paid for the ticket. I came into Lincoln and Mom was in St. Elizabeth Hospital on South Street in Lincoln. Dad was in Lincoln and we stayed at Aunt Bess Huff's (Mom's sister) for a few days. Mom seemed to be better so I came out to Seward to my brother Bob's home and then on out to the farm. I went back to the hospital about every other day.

Dave Huff, (Lincoln, NE) his sister Twila, my cousins, and her boyfriend and I decided to go to the wrestling matches in Lincoln. Dave got a couple neighbor girls to go with us. It was very cold and icy. The car went into a spin and got mashed up some. We went to a place on O St. that made Coney Island hot dog sandwiches, and never did get to the wrestling match.

I stayed one night with Dub Eppenbaugh and a couple of nights with Puzz Menze. Went to Shelby several times with Puzz and Doris. Took one of the Glatter (Shelby, NE) twins home one night and made a date for the next night. When we went to pick her up, her mother met me at the door and said she wasn't home. Figured I got stood up. Later that night we talked to her twin sister and learned that their mother wouldn't let them go with anyone in uniform!

Mom seemed to be improving when I left for Texas. I got into Gainesville and took a bus out to camp. I walked to the 242nd area. It was about 9 PM. No one was there. They were staying out all night on night problems. I went over to the Orderly Room or Sergeant's office. The battery clerk said he was the only one in the battery in camp and the only officer was Captain Karl Salskov, (Idaho), the intelligence officer over at Headquarters Battery. He said they had bad news for me – my mother had passed away.

The clerk and the Captain had another emergency leave all typed out for me and travel arrangements had been made through the Red Cross. I was to go shower and shave and if I needed any extra clothes he would get me new ones out of the supply room. Because the camp buses did not run all night, the Captain would have a cab come in about 2 hours to get me to the bus or train on time. The emergency leave was for 10 days and 4 days travel. They had me on the way back to Nebraska by midnight.

The funeral was the next day after I arrived back in Seward (NE) at the Wood Brother's Chapel on Seward Street. Burial was at Seward's North Cemetery.

There were many relatives, Walsh and Knouse (my mother's family), and neighbors at the funeral. There wasn't room for everyone to get in the Chapel. I remember Aunt Alma and Uncle Lee Walsh, Puzz Menze and Doris Armagost were a great help to me, also my favorite cousin, Geraldine Knouse, one of Uncle Bob Knouse's daughters from Council Bluffs. All five of my Dad's brothers and their wives and some of their children, my cousins were there. Some of the Walsh cousins were in the Service and others had scattered across the nation for jobs due to the war and Depression. The Knouse brothers and sisters and some of their children came except Uncle Ben Knouse who lived in Colorado and Aunt Amy (Knouse) Van Housen in Idaho. Some Knouse cousins had moved away and some were in the Service.

This was the last gathering of the Knouse clan that I know of. It was the last time I ever saw many of them. It was the end of their boisterous, windy, argumentative get-togethers on our farm in D Precinct, Seward County. Our farm had been a summer gathering place for them to eat free, sleep on the floor at night, talk and argue by day. They would play cards and claim someone was cheating. They played poker for matchsticks. They had very little money.

Our farm had always been the gathering place any time of year when there was a funeral of anyone of this very large family. Besides the Knouses, there was also a very large Sperry family, Grandma Knouse's people, and the Lauer family (Grandma Knouse's sister) was also very large.

I never saw my favorite cousin, Geraldine Knouse, again.

I realized at the time I would probably never live on the farm again.

(The Walsh side of my family did a little better at staying in touch. Many of them lived in the surrounding area. After the war, my brother Bob and his wife Sylvia had a dinner for our Dad's birthday that three of his brothers, Matt, George, and Tom and their wives attended. Uncle Tom Walsh celebrated their 50th wedding anniversary in York and nearly all relatives who lived close came that day. In 1979 my wife and I organized a Walsh reunion in Seward and cousins came from all over the nation. The only one of the older generation living at that time was Aunt Teresa, Uncle Ralph Walsh's wife and she was able to attend. This reunion coincided with the printing of our book, "Patrick Walsh of Seward".)

At the time of Mom's funeral I went to Shelby a couple times with Puzz and Doris and had dates with a Shelby girl. She wrote to me the rest of the time I was in the Service. She wasn't very old, about 15 when I met her, and she started sounding serious in the letters. I had only seen her 6 or 7 times. Anyway I only saw her once after I got out of the Army and that was at a dance in

David City (NE). I did not call her or have a date with her. I suppose I should have contacted her because she was a nice girl and I was glad to get the letters.

It was time to get back to Camp Howze and more polishing on the training. More AGF tests, more fellowship at the PX and more good meals in the cafés in Gainesville. These were some of the greatest friendships you could ever have. Back to people you can rely on when the chips are all on the table. Back to a one hundred man team, A Battery, and yes, even back to the five hundred man team, the 242nd that in a few short months would be on the front line where there was no room for mistakes.

Men were transferred out at Howze because IQ tests were too low, and for physical reasons. These guys were not sent to combat zones. They might go overseas to places like Iceland, Alaska, Hawaii, or to islands that had been retaken from the Japanese. They did mostly guard duty and military police, road guides and so forth. Some stayed in the States to guard Prisoner of War camps, did kitchen work in Induction Centers, Reception Centers and things like that.

Some of these soldiers had been in the 242nd since it was activated in Camp White, Oregon, in 1942 and were our good friends. I never understood why they were not replaced immediately after basic training. It would have been a lot easier on them at that time instead of waiting until we were ready to go overseas. (Ours not to reason why.)

June 21, 1944 we got on a train, destination unknown. We carried our clothes and personal effects. All the trucks, guns, cannon and equipment were left at Howze. Sealed orders carried by the colonel were to be opened at a place named on the envelope.

The train headed east, New York, maybe? Then it headed south, - - New Orleans? Florida? Then it turned north again and went through city after city, Washington DC, Philadelphia, East Orange, New Jersey, New York City, and finally stopped at Camp Shanks, New York.

We went through re-checks on clothing and were issued musette bags, A bags, B bags, blankets, gun belts, shoulder harness, etc.

Personnel records were checked. I took out more life insurance. Allotments for wives were arranged.

Classes were held on how to board a ship, how to get into lifeboats, and how to behave onboard a ship.

There were passes to go to New York City. It was a 45-minute train ride, and then a ferryboat ride past the Statue of Liberty finally docking about 2 blocks from the main part of the city. I went into New York City twice. I went to Jack Dempsey's Bar, went to the top of the Empire State Building, to the Hotel Astor where Harry James' orchestra was playing. We went to the Stage Door Canteen and had coffee and doughnuts. We walked into Central Park and generally saw the sights of downtown New York City.

All was in readiness to go overseas. We left Camp Shanks, went to the Brooklyn docks and unloaded inside a large building where Red Cross girls gave us coffee and doughnuts. The doughnut bakers must have made some money during the war! We formed up by battery by rank alphabetically and marched out to the ship.

It was July 1, 1944 when we went up the gang plank of the USS Thomas Barry, formerly the USS Orienta a luxury liner. With the ship's crew, there would be over 5000 aboard.

I had difficulty getting up the gangplank, as did others. We were packing over 100 lbs of gear. My friend, Sgt. Bill Powell grabbed one of my bags halfway up and carried it. He carried only a pistol and I had a rifle so he was not weighted down quite as bad.

At the top of the gangplank stood an officer who called out my name. I answered with "Corporal" followed by my serial number and stepped onto the ship.

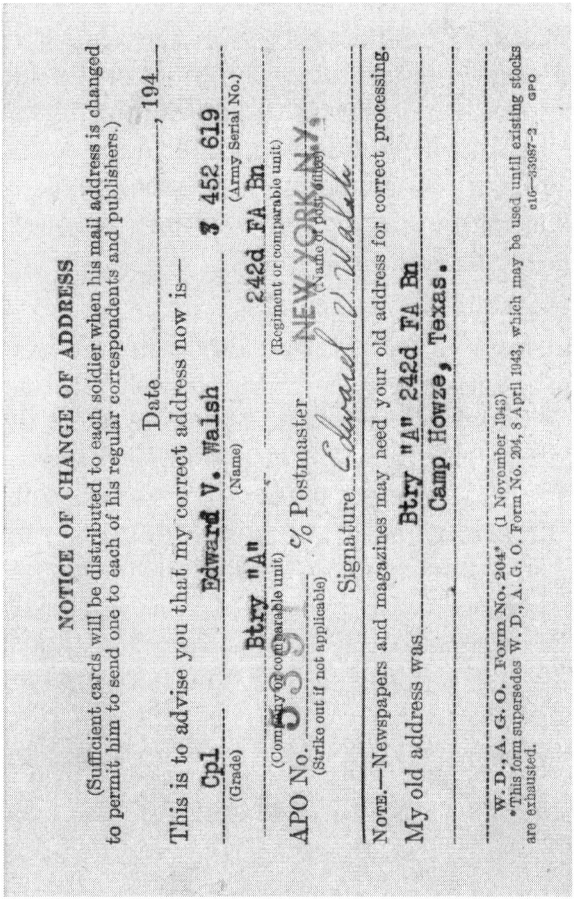

NOTICE OF CHANGE OF ADDRESS

(Sufficient cards will be distributed to each soldier when his mail address is changed to permit him to send one to each of his regular correspondents and publishers.)

Date ____________, 194__

This is to advise you that my correct address now is—

Cpl (Grade) Edward V. Walsh (Name) 3 452 619 (Army Serial No.)

Btry "A" (Company or comparable unit) 242d FA Bn (Regiment or comparable unit)

APO No. 539 (Strike out if not applicable) c/o Postmaster NEW YORK N.Y. (Name of post office)

Signature Edward V. Walsh

NOTE.—Newspapers and magazines may need your old address for correct processing.

My old address was Btry "A" 242d FA Bn

Camp Howze, Texas.

W. D., A. G. O. Form No. 204* (1 November 1943)

*This form supersedes W. D., A. G. O. Form No. 204, 8 April 1943, which may be used until existing stocks are exhausted.

16—33987-2 GPO

A Change of Address

June 1944

CHAPTER EIGHT

We pulled away from the dock about 2 AM, July 2, 1944 in the fog.

Out on the Atlantic there were ships everywhere. This was the biggest convoy ever to cross the Atlantic. Ships had gathered from ports all along the east coast. As the fog lifted and the sun came up, I could see ships for miles ahead and to the rear, to the right and left, just hundreds of ships.

All around the outside of the convoy were Navy ships, mostly destroyers carrying depth charges for enemy submarines. A couple of times during the crossing there were orders to close all water tight doors and we could hear depth charges in the distance. One day there was anti-aircraft fire. Never knew if either were for real or only practice.

We bunked two or three decks down in the ship, just below the water line. A bad place in case of torpedo hits, but there were men below us yet, at least one or maybe 2 more decks below, clear down in the lowest framework of the ship. Everything had been stripped out as far as rooms or walls. All that was left was the framework and sides of the ship with bunks, but mostly hammocks hanging from the super structure of the ship. We could hardly get around because there really weren't any aisles. Just packed with hammocks. It was difficult to get to my assigned hammock carrying all my equipment.

We all had trouble trying to sleep in a hammock. Someone was constantly falling out when they tried to turn over in their sleep.

To the front of these rooms were toilets and showers or "Heads" as the Navy guys called them. The only water in them was seawater and it was usually cold. You could hardly wash your face or shave because it was so salty. A shower was the worst feeling thing you could ever imagine. For drinking water there were places to fill your canteen.

The ventilation was not the best down there, not hot, just kind of clammy. Also the lighting was not the best. Most everyone had brought along something to read and traded with other guys because we spent most of our time in the hammock.

We could only go up on deck at certain times to exercise a little and get some fresh air. There was only a small area in the open deck you could go to on your own and this was also only at certain times.

There were usually quite a few card games going, some for money. They had to put a blanket on the floor and sit and play on the floor. For a few days there was a dice game going on in the corner of the toilet until one guy wound up with all the money. Some of the sailors in the crew were always in these games and knew where other games were on this huge ship. They would smuggle the big winners of the games to another place and there they would play for really high stakes, like several hundred dollars on one roll of the dice. Finally two, three guys wound up with all the dice rollers' money and the dice games were over. A guy from my battery from Louisiana was one of the big winners. When we got to England he sent the money home. (He was the same guy who had been winning every month back in the States, or Stateside as it was called.) He also won the few weeks in England, and the one day crossing the Channel from the sailors. He was the best I ever watched and I saw a lot of them in the service.

There were drills on lifejackets, drills on abandoning ships, drills about lifeboats, and what to try to take with you, and so forth. There were fire drills, air attack drills, and submarine attack drills, closing watertight doors and so forth.

Different outfits took turns going to eat in the different kitchens on the ship. The Navy guys called it the galley.

Even though the ocean was pretty calm, many were slightly seasick. Some were in really bad shape, throwing up and could not keep food or water down. Some had to be carried off of the ship by stretcher when we got to Scotland. I imagine they were really dehydrated by the time we got on land. I had very little trouble on the trip over, just a slight motion sickness for a couple of days.

We enjoyed watching the porpoises swimming right along side the ship. There were hundreds of them going just as fast as the ship just below the surface of the water.

I never knew what other units were on the ship.

We had three or four officers that the Colonel told not to go near the ship's rail. He thought it likely there were some enlisted men who might push them overboard because the officers had been so nasty to the guys under them in the States!

On July 12 we were in the Irish Sea and pulled into the Firth of Clyde in Scotland. We could see Ireland just to the west.

We unloaded at Grenock, Scotland. There was a bagpipe playing to welcome us with the Scots dressed in kilts. Then we boarded a train.

We saw our first signs of war. Bomb damage and piles of dud bombs stacked around. We went on to Glasgow, Scotland, more damage, then south to Carlisle, Lancaster Preston and to Manchester that had much bomb damage. We saw Coventry, completely bombed out, then Worcester and finally Newport, South Wales, where we loaded onto trucks. We were driven to a bunch of tents in a pasture at AberGavenny, South Wales. It was called "Camp Blackbirds." The train trip had taken 24 hours.

We went on long hikes to keep in shape. We would hike through wooded areas that had giant old trees that looked much like the live oaks of Texas. Someone would always start talking about Robin Hood and his merry men, being up in the trees. Some claimed we were in Sherwood Forest, but it was actually 60 or 70 miles away. There was lots of talking, joking and singing, usually something snappy like MacNamara's Band.

At the camp there were showers made out of canvas siding but the water was usually cold. We shaved in our helmets. These were to be the last showers for a long time.

The toilets were a row of holes in boards inside canvas walls and under each hole was a very large pail. Every day local farmers came with a wooden tank pulled by two big draft horses. They dumped the pails in the tank and hauled the contents away. When they reached home with the load, they would add water, stir it up and spread it on the fields as fertilizer. This wagon was called the 'honey wagon.' What hadn't turned to liquid and didn't come out of the spigots of the wagon was shoveled into a compost pile to be used later for fertilizer.

These compost piles were really manure piles and were very large. The farmers had built large cement pits with a grating across the top. Manure was piled on the grating and the high annual rainfall filtered through this. Liquid that collected in the pit was pumped out and spread over the fields. I never did see them use any of the solid waste on the fields. I believe it was left to deteriorate, and the liquid was all they used.

South Wales is far enough north that there is about 18 or 19 hours of daylight in summer. When we got off duty at about 5 PM everyone except the guards took off on foot to the bars. There were not many places to eat as food was rationed. A few places had fish and chips. Everyone boozed it up quite a bit. I usually went to the Irishman's Club for an hour or so and talked to the old Irishmen.

About 8 or 10 of us decided one night to walk a different direction and see if there were any bars or places to eat that way. About 5:30 we started across the fields to where we thought a village might be. After about a mile or so, we came to a field of wheat that had just been cut that day with a binder, the bundles laying in nice neat rows. Five of us were from Nebraska so we said, "Let's give this old farmer a hand" ---even though he wasn't around.

We shocked the wheat in nice rows of shocks. Took us about an hour, maybe two. We never knew if the farmer appreciated having his wheat shocked Nebraska style, or if it made him mad, but we had fun and it gave us a touch of home.

We continued on and found a few pubs and some drinks and talked to some of the natives, mostly Welch, some Irish. Decided we were at least six miles from camp and it was after eleven and getting dark. We bought some booze to take along and walked back to camp in the moonlight. Got back about 2:00 AM, it was already getting light in the east.

There had been large flights of British Lancaster bombers going toward Germany earlier that night. When they went, they were in tight V formation, and now some were coming back in pairs, or singles. Some of the engines didn't seem to sound right.

Another night four or five of us went in another direction and after about 3 miles we found a few nice pubs. We always asked locals we met on the road where pubs, stores, etc, were. They would give directions and we would ask, "How far?" They would always give the same answer, "Only about a 10 minute walk." We soon figured out it would be at least a half hour or more. Now we knew they didn't walk that much faster than us. We never figured out if they didn't really know because they were laid back and had never checked out how far to someplace- - - or maybe they said it to confuse or gripe us. We knew they called us "crazy Americans," or "crazy Yanks."

That night we met a girl on the road and asked what was further down the road. She told us there was a kind of a fair or celebration at such and such place, "only about a ten minute walk." We decided to see if we could find it. About half an hour later we heard music and headed toward it. It was kind of a carnival type thing. There was singing and dancing and booths selling things, probably less than two hundred people in the crowd.

One booth had fish and chips and we really filled up on this. A couple of guys danced a time or two. We bought a few junk things at the booths and listened to the music. I believe I first heard the song "Lambert Walk" here. We had a nice time; the people were friendly and talkative. We headed back to camp, only about five miles this time.

Drivers, assistant drivers, and officers were hauled over to Cardiff one day and you can't imagine the equipment, supplies and the activity around this port area. We were issued all our trucks, Jeeps, weapons carriers, trailers, 105mm cannon and shells. All brand new! We formed a convoy and headed back to the pasture at AberGravenny well spaced because the weather was nice and there was a possibility of air attack. Everything went well and the trucks and cannon were camouflaged under the trees that surrounded the pasture. More guard posts had to be set up to include this perimeter.

A day or so later we went to a firing range along the coast and test fired and adjusted the sights and aiming devices on the cannon. This was done by individual cannon, then by battery four cannon and battalion of cannon, which was 12.

We also had practice bore sighting so if aiming devices should be damaged the cannon could still be used for direct fire. A cross hair gadget is put across the breech of the gun and the gunner finds the target by looking right through the barrel of the gun. The cross hairs are jerked away; the shell is put in, breech closed and the gun fired. Usually all 7 powder charges are put in the shell casing to get the shell to target, like firing a rifle, not lobbing as a Howitzer usually does. On the old Civil War cannon there wasn't any breech to open so the gunner sighted along the top of the cannon.

I never saw this type of firing done in actual combat, but there probably was some in the close engagement battles. I did hear of it being done in the Pacific Theater. Kenny Walsh (Gresham, NE), my cousin, told of his outfit aiming this way at a Japanese command car that had a bunch of officers in and around it having a meeting. One shell got them all. It hit the car dead center.

Back at camp we cleaned the equipment, ate supper and off to the pubs. Over in Europe about this time, Patton's tanks were racing up the peninsula toward Cherbourg. The great tank battle of Falaise Gap had started and we would soon be across the Channel.

(I remember an article in the Star & Stripes newspaper by a reporter telling about the radio conversations from Patton in his Jeep with the guys in the tanks telling them in very calm tones what to do, and the terrible noise of combat in the background. Patton was right there with them at Falaise. He got a little more careful as time went on, but had a lot of close calls all through the war.)

A couple of us were at the Irishman's Club one night talking to a couple of old fellows who had been in France in the First World War. It was interesting to hear them talk in their Irish brogue. They told how terrible it had been in the trenches. Not only the German artillery barrages and infantry attacks, but also the weather, mud, trench foot and flu. They couldn't keep dry, especially their feet. There was no place to go to get warm or dried out. If they could find a little wood to build a fire down in the bunkers, smoke would show above ground and they were immediately shelled by the Germans. Their outfit had been made up of mostly local guys, guys they knew well, and before the war ended they were nearly all killed or badly wounded. One of the old men got so emotional he got up and left the pub. We told the other one we had to go and we left and went to another pub.

I often went to a big pub I believe was called the "The Green Lion," One night we went there and I met a girl who was a corporal in the British Army or Air Force. We had drinks until closing time and I asked if she would be there the next night. She said, "Yes, I will, but you won't. You are leaving for France in the morning."

There were lots of GIs from the 242nd in the pub. Girls went around kissing them and saying "Bon Voyage", "Happy Landings", "Good luck, Yank", "God be with you", and other things. When this started I hung onto the girl I had been talking to and had about 15 minutes of kissing! I never knew her name or she mine and I never got back to South Wales to go there again. As the French say, "C'est le guerre." She was a good looking girl.

So on August 19th, we loaded up everything and took off in close convoy 20 to 25 yards apart. Not the usual 50 to 100 yards because it was raining, the clouds were low and there was no danger of hostile aircraft.

We arrived in Southampton, England before dark. It was still raining.

This was an assembly area with tarpaper huts. Most of the cots had alpaca blankets; real longhaired wool and they were really warm. I would have taken one when we left the next morning but I had all I could carry anyway. Several guys did take them and they were great to have in France.

On August 20th / 21st we loaded on the liberty ship USS Charles M. Hall. Before we went on board, Col. Brownlow came and told some of us the ship was kind of dirty. It seems the last trip across the Channel it had taken a load of "Frogs". I thought, "What are they doing? A big ship full of frogs?" Then it dawned on me that he meant Frenchmen, French soldiers.

Next morning we left the pier but as we got out in the Channel, a storm came up. We returned to the harbor and dropped anchor for the night.

Next morning we weighed anchor and sailed out into the Channel. As we were leaving port the ship in front of us blew up. We went on by. We never learned how many were killed but it looked pretty bad with much damage to the ship.

August 24, 1944 – on to a land of sorrow and terrible destruction. We sailed almost south from South Hampton. The sky was clear as we slanted down the coast. I remembering seeing a lone square house on

Omaha Beach. I have since seen it in several war movies. We came to Utah Beach and dropped anchor about ½ mile out. Landing craft were running back and forth unloading ships. Everywhere there were LCIs (Landing Craft Infantry), LCTs (Landing Craft Tanks), and "ducks" and smaller boats. There were twisted sunken ships and equipment.

When the tide went out there were shoes, clothes, gas masks, and other personal things scattered on the beach. D-Day had been 2½ months before.

Here we were – ready to jump into a raging war. A few will die; a few will be wounded. They will all be scared as hell many times. They will see things that no one should ever have to see. They will lie down to sleep on muddy ground and on snow-covered ground. They will bathe in a little water in a helmet. They will hate the Germans and they will develop a strong dislike for the English and the French. There will be a few from each battery transferred out to fill the ranks of armor and infantry that have had heavy losses in killed and seriously wounded.

"How long will we be here?" I wondered as we prepared to leave the ship.

CHAPTER NINE

We went over the side of the ship carrying one bag and a carbine – down a net rope ladder into a Landing Craft Infantry (LCI). When it was full of men it took off for the beach. As quick as it hit sand under the water, it stopped and the front end of it dropped down. We walked off into the water about hip deep for a few yards, then knee deep, then ankle deep and we were on the beach. We quickly loaded onto trucks and went up the cliff.

The cliff had been cut down to make a steep road for trucks, Jeeps, and tanks. After the thousands of vehicles that had used the road the past two months, it was well packed and solid.

Above the cliffs was the town of St. Mere Eglise. The countryside was green and fairly level with hedge rows running everywhere. About a mile or two in, we paused in a pasture to wait until the rest of the battalion unloaded the next morning.

We had our kitchen trucks and most of the other trucks and the cannon. The cooks set up the kitchen so we could eat and we set up the pup tents to sleep in. The ground was rough from so many shell and bomb holes, foxholes and slit trenches. It was like it had been badly plowed. I had to level off a spot to put a pup tent so the bedroll would be level.

The cooks started to dig a pit to put the garbage in and dug into a dead German. They got the Graves Registration people to come get him to be buried properly in a cemetery and try to identify him.

Next day the rest of the men and equipment were unloaded and we rested the remainder of the day.

St. Lo had been taken after a terrible battle. Gen. George Patton's Third Armor had broken through and taken the Brittany Peninsula and left the cities of Brest and Cherbourg surrounded. He headed for Paris and could have taken it but politics wanted the 2nd French Armor to take it so Patton bypassed Paris and headed for the Rhine River leaving pockets of tens of thousands of German soldiers cut off from getting back to their homeland.

We thought, and most of us hoped, that we might not ever catch up with the 3rd Armor and the war would soon be over. It would have been, but they could not get enough gas to them fast enough to keep the tanks going.

More and more 3rd Army was landing at Utah Beach like us.

Next day we started moving fast. Then we came to St. Lo, worst mess and smell and still burning. On to Avranches, going through Courtances, Grandville, then Bourgeres, Mayenne, Alencon, LeManns, and Chateau Dunn. We made side trips into some of the towns to clean up pockets of Germans but we were always too late. We drove by day and camped in pastures and orchards at night.

Then the infantry and supplies could not keep up with Patton's armored divisions. Our cannon and most of the men set up in a pasture to protect themselves, all the trucks, drivers and assistant drivers were taken to haul gas, ammunition and infantry to catch up with the Armor and to contain the pockets of Germans that were bypassed.

I was an assistant driver so I went along. The fighting at night looked like a distant thunderstorm in Nebraska. Continuous red flashing like lightning with the noise just barely audible.

We hauled mostly infantry and small arms ammunition day and night for a few days. We drove, slept, and drove. We would drive up to where artillery was firing constantly and the infantry would have to walk the rest of the way to the front. We ate K rations. The ammunition was unloaded and we turned back to haul up more. The roads were jammed. Burned out tanks and trucks were just pushed off the road, still burning.

I believe this infantry and ammo went to the area where pockets were left from the Battle of Falaise Gap. This may have been a bigger fight than the Battle of the Bulge because we did not have that many tanks there at first. The rule of thumb was that it took from 4 to 5 American tanks for every German Tank. The British odds were even worse than that.

This transporting of troops went on for a few days and then I was back to the battalion and we were on the move again. We passed through many towns and villages, Orleans, Nemours, Epinal, Chalons, Troyes, Joinville, and Neufchateau. We finally got orders to go into position near the city of Charmes that was full of Germans.

We dug in the guns in a gully below a hill, dug foxholes and trenches. The batteries were almost in a line about a hundred yards apart. Forward observers went up the hill and crawled to the top. I went along as I was in the Survey Section.

We peeked over the hill and there were thousands of Germans down there. The Moselle River ran through the town. The Germans had set up roadblocks and probably had mines on the roads and bridges. There were several tanks to be seen, probably more camouflaged. We saw a lot of horse drawn artillery. Men were sleeping in the sun, some fooling around with a soccer ball, some on guard posts, everyone acting normal, not a bit of excitement to be seen, yet we were sure they knew we were there only about one mile from them. They could have wiped us out in less than an hour if they had wanted to.

We called Group Headquarters and they said, “Get the hell out of there.” And we got out of there fast.

First taste of combat: September 14, 1944 - - We pulled back to a village, Mirecourt. I did not go to Headquarters because it was about dark and I didn’t know where they were.

At dark the Germans started to move to try to get out of a pocket they were in. Our whole battalion started firing, fired most of the night on the roads and crossroads the Germans were using to escape.

Sgt. Reed and I sat down by a tree about 25 yards behind the cannon that we were firing continually. We talked about how bad it was going to get. He said he guessed this was what he had been in the Army for all these years. He said, “What do you think, kid?”

I said, “It’s a little like Fort Sill, these first few hours.”

A little after midnight he said, “I’m getting sleepy. I guess I’ll take a nap.”

I said, “Me, too.” We leaned back against the tree and slept for about 3 hours.

When we woke up, the guns were still firing. All you could see was the dull red flashes of the cannon, the smoke around the flash and the silhouette of the cannoneers. Eerie.

Most everyone got so they could sleep while only a few yards from the terrible noise of the cannon. Learned it in Fort Sill, I guess.

We had some of the 79th Division in front of us and the 310 Field Artillery Battalion behind us. The Germans counter attacked a few times but the breaches in the line were repaired by our artillery fire.

The next morning there were many German prisoners, many dead Germans and all kinds of ruined German equipment. There were also lots of dead horses. There had been quite a bit of horse-drawn artillery. There were many wounded, both German and American and quite a few Americans killed. We now knew we were fighting a German Reinforced Combat Team, probably 2500 to 3000 men. We got darn near all of them, maybe 2 or 3 hundred escaped.

On the 15th of September we moved to a position back near Charmes, still in support of the 79th Division. We fired for them several days until they got a bridgehead across the Moselle River. We crossed the river and continued to fire for the 314th Infantry Regiment of the 79th (over 1000 rounds) to support them as they crossed the Meurthe River. Then we all moved to positions north of Luneville.

One night a slow moving bed check Charlie plane came over us, and everyone started shooting and he was shot down. We caught hell for this because you were not to shoot at aircraft at night because it might be one of ours.

Early in combat we were strafed by a ME109, which scared everyone pretty bad. We had just gone into position and had not had time to dig foxholes. I almost jumped in a creek and it's a good thing I didn't. What I thought was a little creek turned out to be about 8 feet deep and I can't swim.

During this time the story went around that an infantryman was wounded between the American and German lines. Two of our medics put up white flags and a Red Cross flag to go get him. The Germans quit firing. The medics got him on a stretcher and started carrying him back to our lines. The Germans shot and killed both the medics and the wounded guy they were carrying.

The American infantry in this small area never took a German prisoner for over a week. They killed every German they captured. You never hear things like this about our GIs but I knew two different guys who saw it.

We were in the position by the creek for a few days. I saw Bing Crosby go by going up to entertain the infantry ahead of us. It was rainy and foggy. He was up there less than an hour and over the radio the forward observers called for a fire mission – that three German tiger tanks had broken through the line and the infantry was backing up. We fired quite a bit, got one of the tanks and the other two pulled back. Right after the call came to us, they brought Bing back to the rear right by us again and they were driving fast. The infantry said he had only sang two songs before the tanks broke through.

Once we were bogged down by many Germans ahead of us about 3 or 4 miles in a fair-sized town and air support was brought in to help. It was a flight of about 24 medium sized bombers, B25 or B26. When they made their drop, there were terrific flashes of fire and in a few seconds the noise, and then a flash of heat. And we were that far away! What must the noise, flash and heat have been from the big bombers over cities?

Everyone was always salvaging something from crashed German planes that had been shot down, like machine guns, plexiglass, wire, etc.

We had forced a bridgehead at Charmes and were in a town somewhere close to the east side of the Moselle River. It was a bright moonlight night and we were standing guard out in the street. Suddenly there was a big red flash to the east of us quite a way behind the German lines. Just a few seconds later, something that sounded like a freight train passed overhead. A little later a big red flash followed by a boom way back behind us.

This went every 18 minutes all night. We soon figured it was a big gun on a railroad car. Probably about a 280 or 340mm, a shell about 12 inches in diameter and 4 or 5 feet long.

There was a sound and flash battalion in the area and they were getting a coordinate on it but we didn't have anything that would shoot that far.

At dawn here come P47 planes from the 9th Tactical Air with bombs under their wings. The Sound and Flash people had given a good enough reading that they found a camouflaged railroad tunnel the gun was in and bombed the tunnel shut. Some said the gun got out the other end, some said both ends were bombed shut. Anyway, we never heard the big gun again.

We found out the Germans were shelling Nancy, France, where Patton's Third Army Headquarters were. They almost got Patton and other big brass there. Some officers were killed and some wounded that night in Nancy as well as civilians.

In this area we saw and heard "Buzz Bombs" and "Screaming Mimis", a thing that made a terrible noise but not too much damage. We saw some flying bombs in northern France and some kind of rockets, V one, or V two.

With armor, more artillery and some of the 79th Infantry Division, and after quite a fight, the Germans had been forced across the Moselle River to a new line on the Muerthe River. We got this done just in time because farther south, the Germans were beating up on the GIs real bad. We came in on the side and a little behind them and relieved the situation at Luneville so the Americans could regroup.

When we came in on the side, the Germans tried to retreat, and many did, back to the Muerthe. The Air Force spotted a line of horse drawn artillery about two miles long going around the base of a good sized hill on a narrow road with a small stream running alongside the road. They dive bombed the road at the front of the column and strafed the whole length of it. They called for artillery and practically all the artillery in the area fired on the column from one end to the other.

It was late in the afternoon, and after dark the few Germans left took what horses were still alive, picked up their dead and wounded and escaped to the Muerthe River line. Next morning when we went down this road, it was filled with broken equipment scattered all over. We had to push dead horses off the road to get through. The Air Force estimated there were 200 men killed or wounded. The pilots were the only ones who saw what happened. It was a mess, but at least they took their dead and wounded out. They had lost practically all their equipment.

In this same area another horse drawn artillery group was seen by the Air Force going down a mountainous road. Just in front of the column about one fourth mile was a bridge. They bombed the bridge, blew it out and bombed the rear of the column. Then they strafed it and called for artillery to finish the job.

Few, if any, escaped because infantry got in there to take prisoners. A total of 2000 were killed, wounded or taken prisoner, but a surprising few of around 175 actually killed. (It is estimated that at least 70% of the casualties in WWII in combat were caused by artillery. There were many casualties caused by aerial bombing but these were mostly civilian.)

About the end of September we went into position about a mile from the Meurthe River with the Germans dug in behind a road that ran parallel to the river about half a mile on the other side. The Howitzers started firing immediately. The rest of us were starting to dig foxholes.

Then everyone heard in-coming artillery. There was a large culvert under the road nearby. I ran with some others to crawl into the culvert. Someone yelled, "There's a dead German in here!"

Someone else said, "Get him out, we can't all get in there with him in there."

Several said, "Well, I'm not moving him."

Finally a couple guys drug him out. By that time the shelling had stopped. We figured out the shelling had not been that close. We went back to digging foxholes. I got mine down about a foot, when a march order to cross the river came.

The infantry had made a bridgehead and were across the river chasing the Germans. It was close to dark by now and we moved across the river to the position the Germans had occupied.

There were dead Germans, burning and wrecked equipment everywhere along the river on both sides. Not many Americans.

When we got to the road, it was even worse. There were dead German soldiers all up and down the road, under the trees and out in the fields. At the point where the road from the river came to a T with the parallel road, there was a very large house made of stone and wood, a house, barn, machine shed combination all built in one. This structure was badly damaged and burning in several places. There were dead Germans, some dead civilians, also wounded Germans and civilians. There were quite a few head of livestock dead and wounded. None of the dead were picked up that night.

The forward observer who had called in the coordinates that resulted in this carnage was Captain Carlson. He had directed the fire up and down the river and the road. He had gotten the idea of using a quick fuse on the shells. They would hit in the top of the trees and go off like time fire about 20 or 30 feet in the air. This had eliminated about everything in the area long the road and along the river. Then he went to delay fuse and started on the big house. The delay fuse would penetrate the house and go off inside.

As we all looked at the ruins of the battle, Captain Carlson walked up the road and along the river about ¾ mile each side of the house and looked over this terrible carnage.

Everyone in the battalion heard about the house and nicknamed it "Carlson's Castle.

The C Battery agent and I walked around looking for a good place to bed down. Right in the middle of a bunch of dead German soldiers, he said, "The further we go up this road there seems to be more of them. We might as well pitch our pup tent here, don't you think?"

I didn't know if he was joking or not but I thought, "You're not going to get the best of me." So I said, "Fine with me," and laid down my shelter half, blankets, and musette bag.

We pitched our tent with bodies only about 20 feet away. Went to eat, came back, went to bed and slept real good until about 4:30 AM when it started to pour down rain and kept it up until daylight. Then we got up, packed up and went to breakfast. The dead ones had turned a blue color overnight with a slight tinge of red in places on their faces.

Finally about 8 AM, graves registration trucks came and picked up the dead Germans. One guy would get them by the feet and another by the shoulders and start swinging. They would say, "One, two, Heil Hitler" and give them a toss in the back of an open 6x6 truck until they had a load. They did not bother to rack them like cordwood as the Germans did at the concentration camps, just tossed them in a pile.

In two or three positions around here we fired over one thousand rounds getting the 314th Infantry regiment of the 79th Division across the river.

Talavera's twin brother was in the infantry of the 79th and Talavera tried to contact him during this time but did not succeed.

March Orders – Move Out

Combat 1944 - 45

German horse drawn column strafed and shelled.

Note some horses are still alive.

Combat 1944 - 45

Shelled out village

Combat 1944 - 45

CHATPER TEN

September 27 – October 14th, 1944 – Ferret de Parroy.

The infantry was bogged down and could not move. The Germans and our infantry were fighting in this heavily wooded and brushy forest (Ferret de Parroy). The trees were so tall that no sunshine came through to the ground. We could not use tanks or artillery in here and our infantry could not dislodge the Germans.

We did have forward observers, radio and telephone in there in case we broke through to the other side of the trees and we could use artillery. We were in this spot for days and could not move.

Finally one day, as 2 other guys and I were standing road guard, several Jeeps came down the road. General Eisenhower was in the first one, General Patton in the 2nd and General Patch in the 3rd. There were several one star generals, possibly Division Artillery commanders, and two 2 star generals. They got about half a mile down the road toward the Forest of Parroy when the Germans started shelling all along the roads and crossroads in the area. They evidently knew the Big Brass were there and they hoped to get them.

We hit our foxholes in a hurry, but the closest shell was several hundred yards away. There were three of us at the guard post, Corporal Adolph Zazadil from C Battery, Corporal Herbert Emerton from Headquarters Battery and me. Zazadil was from Chicago Illinois and Emerton was from Taylor, NE.

The generals got thru okay and as far as I heard, no one was killed in that barrage.

From the 27th of September until the 14th of October we fired for the 79th Division in the battle for the Ferret de Parroy. Here practically all officers and men had to do some forward observing. The things we learned in the States about being able to take over any man's job paid off.

Walsh and Maurice Wagner, Charles City, IA

At the Forest of Parroy.

The night of September 28, Lt. Allen Burton, Chester, Connecticut was a forward observer and was killed by a German patrol. We found him stripped of all his clothes, even his dog tags. Probably wanted to use his uniform and dog tags for an English speaking German to infiltrate our lines.

Walsh, Adolph Zazadil, Chicago, Illinois, Herbert Emerton, Taylor, NE

At the Forest of Parroy

A few days later First Lt. Thomas Wilson, Lincoln, Kansas was severely wounded and was evacuated to the States. Then in a couple of days Sergeant Charles

Anthony (Lexington, NE) was severely wounded and was evacuated.

Finally the small forest was shelled heavily and bombed and strafed by our planes and the Germans retreated. It was a terrible mess in that forest.

On the 14th of October we displaced to the vicinity of Fort Mannonviller. Again we bogged down and could not advance. This ancient fort, (I don't know how old, but it was old enough to have a moat) was on a hill overlooking a valley. Behind it was another small valley. Our three firing batteries were down in a pasture about one thousand yards apart in a slight inverted V shape.

Behind our batteries was the 999th Field Artillery, which was a black battalion with white officers. They had 155mm long barreled rifle-like artillery pieces called "Long Toms", pulled by caterpillar tractors. This tractor was geared higher than a regular caterpillar and had a little armor on it. They were a big help to us on targets we could not reach and also when our trucks or Jeeps got stuck in the mud, they would pull us out with their caterpillars.

Each battery had several single 50 cal. machine guns around itself for guard duty and anti-aircraft use. Each battery also had a 50 cal. machine gun mounted on a circular track above the cab of one truck, plus a 30 cal. carbine or a 45 cal. Colt pistol or a 45 cal. machine pistol, plus a few bazookas, plus Browning automatic machine guns. An anti-aircraft battalion was on the perimeter of our battalion and the black battalion.

At the peak of the hill was a road going into the old fort off a regular road and we always had a guard there, sometimes called guard, sometimes called road guide. Actually both to guard and to give information about finding someone or some outfit to anyone coming along the road.

Well, if you are standing up there, the Germans could see you, so the Colonel said be sure to dig a foxhole there. (He always said that at every road guard place.) This time he said he knew damn well the foxhole was not always dug, but we better dig one here. So whoever stood guard there the first time had dug one.

I came out that afternoon for my 4 hours of guard and it was bright sunshine out. I didn't think the hole was quite deep enough so started digging in it. It had been raining all the time so the ground was soft and muddy.

Someone stopped for directions and I told them, and started back to the hole. Just then an 88 whizzed in and hit about 15 yards behind me. I dived for the hole but before I got there, another hit a little to the front. Plastered me with mud but no fragments hit me. Lucky both shells went in the mud and most of the force and shrapnel went up instead of sideways or I might have been cut in two.

I was in the hole by now and waited a minute or two to see if there would be more but they must have thought they got me. I looked out and all of our telephone lines that had been laid on the ground were blown to pieces.

I crawled the hundred yards or so to the fort and told the Colonel the lines were out.

He said, "What the hell happened to you? You are white as a sheet and how did you get so damn muddy?"

I told him about the shells hitting near me and he said, "Well, did you people dig your hole. You could have got in it."

I said I didn't have time.

He said, "Well, what if they would have sent in five or six more rounds, would you have been glad the hole was there?"

I said "Yes, sir!"

He said, "Well, don't stand here. Get the wire and telephone people and get out there and show them where the break is and tell them to stay on their bellies while fixing it and you, you stay in the hole."

The wire was fixed and there was no more firing. The Germans must have thought they got me or that one guy was not worth the trouble.

The batteries below the hill were strafed by an ME109. They cut loose with the anti-craft and 50 cal. machine guns and got him. He was burning as he came over the hill. He parachuted out between us and the German lines and the infantry captured him. They brought him back to our headquarters.

The infantry had taken his sheep lined flight jacket and pants, and his sheep lined helmet and boots. They brought him in, in just his uniform and stocking feet. It was very cold out.

The Colonel said he would court martial the guy that did it if he could find him. The ones who brought the German in just laughed, so the Colonel got in touch with their officer by radio and the officer laughed at him, too. The Colonel really got mad and wanted to know the officer's name and serial number. The officer said if he wanted to find out he should crawl out to where the infantry was and see what he could do about it, that maybe they would take the Colonel's coat and boots too if they were dry and warmer than what the infantry had.

We put some blankets around the pilot and took him back several miles to Service battery and turned him in as a prisoner of war.

Down in the rooms and hallways of the old fort, people gathered around in groups where they slept or read books or played cards. It was dark, even in the daytime, so they took any bottle they could find, and filled it with gasoline they stole out of the truck tanks. This was one hundred octane aviation gas. They put a long rag wick down in the bottle and set fire to the rag. It made a good light to read by and gave some warmth. But it smoked such black smoke and lead fumes, that everyone's clothes, face and hands were black after a few hours of being around them. No one cared, or was afraid of them. We did not know about lead poisoning, clean air, smog, etc. I never heard of any of these lights blowing up, or breaking, or causing a fire.

Of course this is the same thing as a "Molotov cocktail." If thrown at a tank or truck, it would explode and cause a terrible fire for a few minutes.

A Battery's position below the Fort was shelled, and we had to move to an alternate position (which was already picked out ahead of time in case you did come under fire.) When a battery gets shelled, it is called "C.B. or Counter Battery.)

The fort had had a moat around it and parapets around the top, many rooms of different sizes in it. It may have been built higher originally, we could not tell because much of it had fallen to ruin.

We could see it had been in existence a long time and been attacked by some old cannon that fired solid balls as we found some of them when digging latrines and garbage pits at the bottom of the moat. We also found arrowheads, spearheads, old pottery and pewter ware.

A couple of us crawled out on a parapet with a forward observer and watched the German lines with field glasses. We could see some activity, one, two, three Germans here and there, but not enough to have the artillery fire on them. Once in a while the Germans would bounce a shell off the fort because they knew there were a couple hundred of us in there and that observers were watching them – just enough to keep us under cover.

Fort Mannonviler, Combat 1944

Another agent and I were on guard at the entry to the fort one night. It was cold and pouring down rain again. We were standing at a place near where there at one time had been a drawbridge.

Here comes Captain Carlson. The only clothes he had on were his long handle underwear, not even socks. He was soaking wet, covered with mud. He had come up the hill from the firing Battery A. He must have slipped and fallen a number of times. He was muttering and babbling.

I said, "What's wrong, Captain Carlson?" He had been our captain or commanding officer of A Battery ever since we went into the service at Camp White, Oregon. Many times he would ask Talavera and me up to his hotel room to have a drink when we were in town for a weekend. (You could not have a drink in a public place together with an officer.)

He recognized me and said, "Walsh, I killed all those boys. I killed all of those young boys!"

We got a hold of him, got him out of the rain, found a blanket and put it around him and wanted him to go to the medics, but he kept saying, "I killed all those young boys."

About that time Colonel Brownlow came out of one of the rooms, (they had no doors). "Get him in here," he said. So we put him in the Colonel's room.

Someone called the medics and a doctor came. They got the wet long johns off him, put dry ones on and wrapped him in blankets. All the time he kept repeating over and over, "I killed all those boys. Walsh, I killed all those young boys."

The doctor gave him a shot and he finally stopped talking, giving a moan now and then. An ambulance came and took him back to a hospital, back to Paris.

Before we got out of that room, the Colonel said he did not want any talk or rumors about this to get to any of the batteries except A battery, of course they would have to be told. He threatened court martial if anyone got caught talking about it. I don't know if he could have done that or not, anyway, eventually everyone knew about it.

Captain Carlson had had a "nervous breakdown" or as it was called "battle fatigue." So one of our Lieutenants was made commanding officer of A Battery and later would be promoted to captain, Captain Reginald

Carpenter, (South Bend, Indiana), who lasted the rest of the war.

One of our outfit saw Captain Carlson later in Paris. He had been in the hospital about a month, then a rest center for 2 months. He was then put on duty in a rear echelon outfit of the quartermasters.

We stayed at the fort area from October 14 until November 17. It rained continually the entire month. The trucks could not get ammunition to the guns in the mud so it had to be carried by hand. The trucks would get stuck and had to be winched out, or we would go to the 155 Long Tom Rifles (the black guys) near us and get them to pull us out with their half-track caterpillars that they used to pull their guns. We got pretty friendly with these guys, and also with the battalion of anti-aircraft artillery that surrounded us. At the time we had Counter Battery it was a hell of a mess moving everything out of the mud.

During this time Service Battery got a small rest center going in Luneville and the men could get a 24 hour pass to go back there and take their first hot shower since we had left England, and sleep overnight under a roof. We weren't there long enough for many of us to get to go. I don't think any of the agents got to go.

On October 23rd, the 44th Division relieved the 79th Division Infantry. This outfit was new and had never been in combat. Our few months made us feel like old veterans. The first week or two we worried that the Germans might break through the 44th, but the new GIs held okay.

November 17th we laid out a heavy barrage at dawn, and the 44th began to move. We were right with them, going in and out of positions with extreme difficulty in the mud and rain.

Robert Jackson, Chicago Illinois

Edward Walsh, Gresham, Nebraska

Fort Mannonviller – Combat 1944

As the Germans fell back, Corporal Robert Jackson, B Battery, Chicago Illinois and I walked down the hill from the old Fort on the German side just to look over the area where our infantry and the Germans had faced each other. The little bit of the area we could walk over was all foxholes, and shell holes, with German and American equipment laying around everywhere. We crossed over a fence and darn near stepped on a dead

German laying in the tall dead grass. We decided we had seen enough and went back to the fort.

From November 18th to the 23rd we changed positions sometimes several times a day and also under cover of darkness. This attack carried us into the Vosges Mountains. Some of the positions we held were near Lantrey, Amenoncourt, Fouleray, and St. Beging.

Early one morning came the call "Close Station. March Order." (We were moving up.) Many were still sleeping and, of course, were in their clothes and boots. You never took your clothes off to sleep in combat, and had your weapon right beside you.

Everyone tumbled out, tore down pup tents, loaded everything on the trucks, and hooked the cannon and trailers behind the trucks. In a few minutes all the vehicles were headed out. All three of the firing batteries went down the road in order of march, Abel, Baker, Charley and Headquarters. Service Battery several miles back would move up later. We moved out slower than usual because of the fog.

We pulled into a village after a few miles. It was still raining, and there was still heavy fog. The village and all roads in and out were jammed with trucks, jeeps, tanks, artillery and at least a thousand to fifteen hundred infantrymen. The Germans were still throwing a few shells into the town. Things were so jammed we were moving at a snail's pace.

About 20 or 30 infantrymen were lined up at one building. One of them came out of the building and crossed the street between our trucks. Someone yelled to him, "What the hell are those guys lined up for over there?" The infantryman said, "It's a whore house." Someone asked, "How many whores do they have in there?" The guy answered, "One." Somebody else says, "One? And all those guys?" He says, "I don't know about them but I was number 37 and it seemed all right to me!"

You couldn't believe all the laughing and hollering about this as we continued to crawl through the village.

It was still raining, still foggy and the enemy was still dropping a shell into the town every few minutes. The closest to us was about a block away. A few got hit with shell fragments but we didn't see anyone seriously injured along the street where we were.

If it hadn't been for the fog the Germans would surely have known how jammed up the village was and it would have been wholesale slaughter.

In late fall we were in the Saarbourg area and things had been pretty tough. We did not have much in front of us in the way of infantry, armor, or mechanized cavalry. It was a cloudy dreary day. We had set up the Message Center in a building in the middle of a town; the firing batteries were just outside of town and doing quite a bit of firing.

The main road or street through the village toward the Front was kind of S shaped, a right turn, and then a left turn over about a three block long area. We were about a half block to the right of the left turn.

About midmorning we could hear tanks to the rear of us and knew by the sound they were American. There was a lot of difference in the sound of German and American tanks. The tanks started coming down this street. With them were Jeeps, halftracks, trucks, weapon carriers, truck drawn artillery, self-propelled artillery, and men by the hundreds. They were moving fast thru this narrow street. (This was to continue hour after hour all day and into the night, just unbelievable.)

A little boy about 9 or 10 started riding his bicycle back and forth out in the street. We located his mother and told her to keep him off the street, to put the bike away.

She did, but about an hour later, there he was again riding up and down, back and forth across the street in the midst of all this traffic. He was dangerously close to where all those vehicles were turning. Someone got his mother again and had her get him off the street but it was less than an hour later, there he was again riding up and down the side of this narrow street.

This time he got too close to a turning truck and the rear corner of the truck where the tailgate hinges extend out about six inches caught him in the throat and cut his jugular vein. The bicycle went flying. The driver of the truck never knew this had happened. The stream of traffic continued.

Several of us standing along the street ran to the boy yelling for medics to come and a couple of them got there about the same time we did. There was nothing they could do. He bled to death in minutes. By that time one of the battalion doctors was there but he, too, could do nothing. There was a giant pool of blood in the middle of the street. The mother came out and fainted. We carried the boy's body and the mother to the house. Two different times later in the day the mother passed out in the street.

We sure had our hands full with the mother and the neighbors. Some of the neighbors got hostile and we finally had to search their houses to see if they had any guns. The guards were pretty edgy that night. We usually didn't worry too much about civilians shooting at us or sticking a knife in someone's back in the dark but this was not a normal day and the night would not be normal either.

About an hour before dawn the agents were called to take firing orders to the firing batteries for a dawn barrage. It was black as pitch, cloudy, no moon. The big problem for the agents was to get into their battery's area without getting shot by their own battery guards. You better know the password and countersign and hope whoever was on guard was not trigger happy and started shooting first and ask for the password later.

Anyway, we all got in all right. The guard post I came upon yelled, "Halt! Who goes there?" I said, "Walsh, the battery agent." They said, "Are you alone?" I said, "yes". They said, "We recognize the voice, but what's the password?" I gave it and they said to come on in.

Right away they wanted to know what was going on. I said I had a message for the Captain for a dawn barrage. I said I would tell them more after I had talked to the Captain. I went to his tent and before I opened the flap to go in, I said "Is anyone awake in there?" (You don't dare go in a tent and shake someone's shoulder that was probably sleeping with a gun in his hand.) A voice said, "What the hell's going on, Walsh?" It was Executive Officer Lt. William Maddox (McAlister, Oklahoma) who recognized my voice. I told him I had a message for the Captain unless he was upfront as forward observer, then the message was for the Exec. He said to come on in and he would light a candle. He said, "The Captain is up front," so I handed him the message. It was a list of targets and the time to the second when to fire on each of them.

He said he had figured something big was up with all the stuff that had moved by the day before. I said, "Yes, from what I've seen moving thru this town, I'll bet there's over a thousand pieces of artillery in this area. There's really going to be some fireworks."

By then everyone in the tent had his gun belt, coat and helmet on. I said, "I'm going back to headquarters."

Lt. Maddox told me to watch out for the guards. I said, "I'll stop and talk to them on my way." We stepped out of the tent and the Maddox started yelling, "Fire

Mission! Fire Mission! Everyone out!" He had always been loud but in the quiet black night he was really loud. The gun crews came stumbling out of their tents and ran to the Howitzers. The officer began giving the data for the first target. They would have to wait to fire because it was about 20 minutes until dawn.

When I got to the perimeter guard post, one of the guards said, "What the hell's wrong with Lt. Maddox? He's so damn loud he probably woke up the Germans! Don't he know they are only about a mile and a quarter in front of us?" I said, "Yeah, he really gets wound up over these firing orders. But you got to admit that him and Captain Carlson are the most likeable officers you ever run into." They agreed. I said, "In about 15 minutes or so, you are going to see something you will always remember. There is probably over a thousand artillery pieces in this area and they are all going to be firing continuously for about 45 minutes."

I started back to headquarters; got part way and a tinge of pink began to show thru the clouds in the east. Suddenly the ground shook and the sky lit up like daytime, like being in the worst thunder and lightning storm you could imagine. I could imagine I could hear Lt. Maddox yelling out the firing orders for "A" Battery but surely not. I could hear the shells going off in the German lines, which doubled the noise and the flashes of light.

This continued for not quite 45 minutes and then I heard the clatter of tanks starting to move forward with the infantry. Some of the firing died as tanks went forward to visual targets and the batteries went to targets the forward observers of the individual battalions located.

Slowly the whole front moved forward, but slowly. The main objective was to drive forward to the Rhine River. If this was not accomplished, the second objective was to straighten out the front line and occupy more stable positions before winter.

The first did not happen as resistance was too strong, but we did move forward quite a few miles that day and made the line straight and in better position. As we moved we saw lots of destruction from the early morning firing, buildings burning, and quite a few casualties.

I read later in “Stars and Stripes” (and also in Patton’s book) that there were twelve hundred and sixty cannons of different sizes in that barrage, and probably 40 to 50 thousand men. Just consider the amount of gasoline, ammunition and food it took to supply those in only this 25 to 30 mile front.

The next day, Lt. Maddox said to me, “Well, Walsh, we sure gave the Germans a Texas style show, didn’t we?” (He was from Dallas.)

Serving the Piece – 105mm Howitzer

Firing into Germany – Combat Spring 1945

CHAPTER ELEVEN

One night we pulled into a town and set up the Message Center in a nice two story house after telling the people to get out. Campbell and I found a room that had a big feather bed with two feather ticks, one to lie on and one to cover up with.

We were dead tired so we crawled into the bed, all our clothes on, even boots, with our coats, helmet and gun nearby. We dropped off to sleep right away. In about two hours we both woke up wringing wet with sweat. Those feather ticks were really warm!

(I knew then why a feather tick had been such a great asset to the early settlers in Nebraska. I remember we had one feather tick on our farm in Seward County but we slept on it, not under it.)

The Message Center was part of Headquarters Battery, and was made up of these men: Message Center sergeant, one corporal, one technical corporal, and the four corporals from the batteries, A, B, C, and Service. A guy named Raymond Centers (San Francisco, California) was the Sergeant, Herb Emerton of Taylor, NE, was the Corporal. The Tech Corporal was Ellers. Centers was transferred out during combat and Emerton became the Sergeant.

Now this Message Center did not have too much to do with the Battalion switchboard or the radio in the Colonel's command car. Most of the messages were more secret and were delivered by hand or word of mouth by the four agents from the batteries. The agents also brought in battery statistics every morning. The Message Center combined all information, which was then given to the Colonel. All promotions and write-ups for medals were done by the Center.

The Center helped keep the generator running that made lights for the Fire Direction Center, only about 2 or 3 bulbs. Sometimes we ran a line with one bulb for the Message Center. The Fire Direction guys had to have a fairly decent light to read maps, do overlays and coordinates, read slide rules, etc. At times they used gas lanterns or flashlights, even candles when moving fast and there was no time to set up generators.

For most of combat the battery agents were A – Edward Walsh, B – Robert Jackson, Chicago, Illinois, C – James Campbell, Campbell, Nebraska or Adolph Zazadil, Chicago, Illinois and for Service Battery – Maurice Wagner, Charles City, Iowa.

At Camp White and Yakima I had first been a heavy machine gunner. Then in Fort Sill I became a machine gun corporal, but was actually recording the data of the four 105mm Howitzers of A Battery. Then I was transferred to the Survey Section of A Battery and became Battery Agent. I understood some of the survey, the geometry part of it, but was lost when it got into logarithms, slide rule and high math. Being the agent, I did not need to know this. The agent is a messenger, a runner, a liaison corporal to carry messages, oral, written or coded.

(In trench warfare in World War I, they were called "Parapet Poppers" because of popping out of one trench and running to another trench, from headquarters in the rear up to the front line. Hitler was a famous "Parapet Popper" in that war, was wounded once with shrapnel and gassed once with mustard gas. He was awarded the German Iron Cross for bravery, one of their highest awards.)

The Message Center people and the agents took their turn at guard around the message and fire direction center and the switchboard if used. If there was much firing going on, the Fire Direction Center men did not have time to do guard duty.

In combat each section of 8 or 10 men had guard all the time, not one but two guards day and night. All had their turn at guard, sergeants, corporals, and privates. I guess it was really hairy, cold and wet out around the Howitzers but we also had some bad ones at the message and firing center at headquarters, especially from about 11 PM until daybreak. Some times we could hardly stay awake because we did not walk. We just stood in some spot or doorway, maybe an open window or some place in the shadows and watch, wait, and listen. Your imagination goes wild. You are armed with your regular 30 cal. carbine with extra clips in your ammunition belt, a hand grenade or two, your home made knife and a German Luger you have picked up. Plus at Headquarters, we had a BAR, Browning automatic rifle.

We had moved into a strange village just at dark and we had had some shelling problems that day. We were really too close to the front for artillery headquarters. We set up in two or three houses. These houses were also the barns, that is, the house and barn were connected together. The cattle, chickens, horses, rabbits all lived in the same building with a manure pile in front of the front door. This village was in France, but all the people were German, the old Alsace Lorraine-Saar Valley, Rhine valley deal from World War I.

Anyway this time the four battery agents and some other message center guys set up in the barn part of the building. Some of us put our bedrolls on the hay in the barn to sleep while 2 stood guard.

Jackson was to be on guard with me and he was nervous because he had been up front with a forward observer and had gotten hit in about 20 places with small shrapnel. He was scared all the time. I was just about asleep when he says, “Walsh, I can’t sleep here.

These people in this house are German and they will kill us with a pitchfork or something if we go to sleep."

I said they won't bother us, they are scared, too, and they are old. We'll be okay. Besides we have two guys standing guard in the doorway not 20 feet from here.

I guess he did not sleep, but I did.

It came our turn to guard at about 2 in the morning and we got up and stood in the doorway and watched. It was quiet except for an occasional incoming German artillery shell and these were not anywhere near this village.

Suddenly Jackson says, "God, something moved by those bushes across the road!"

I watched and sure enough something did move. Jackson says, "What will we do?"

I said if it's a German patrol we have to stop them from getting any closer. So we said, "Who goes there?" Give the password!"

No answer. We repeated, "Who goes there?" Still no answer, but more movement.

Jackson says, "Let's let them have it", so he cut loose with the Browning Automatic and me with a carbine.

In no time a lot of our guys were around and we crossed the road. We found we had killed a bunch of chickens that had been roosting on the lower branches of the bushes!

I don't think Jackson slept a wink that whole night!

We went into position near another town late one afternoon and I started to walk from A Battery to Headquarters about a mile away in the village. As I went down the road there was a high spot and German artillery saw me and put three rounds of 88mm in by me too close for comfort. I jumped in the ditch and crawled for about a quarter of a mile toward the village, then got up and ran the last half mile in the ditch.

Jackson was on guard in a doorway with me that night and the Germans shelled the town all night. Some were pretty close. Next morning we caught a German in the church tower. He had been directing fire on the town all the time we were there. Our own artillery had spotted theirs and put them on retreat. Jackson was really wild that night. We finally got some sleep the next day.

Another night I was on guard with Campbell (C Battery agent). We had been in this village for some time. Everyone wanted guard duty the hours the village people were going to bed because they went to the toilet on the manure pile in front of their houses. Everyone got a bang out of this when the villagers visited the manure pile before going to bed. You couldn't really see anything, but you could hear ---- and old Campbell would about die laughing.

When the German was in the church tower, the room we had stood guard in was full of rabbit cages. The next day Jackson turned all the rabbits loose in the room and then turned the buck rabbit loose. He thought this was really funny.

Later on the night the chickens were shot a German shell hit our roof, which was tile. The shell blew a big hole in the roof and the tile fell in on all of us sleeping in the loft. We made a run for a house that had a basement. The Message Center sergeant, Sergeant Centers, was in the lead. He came to a manure pile that was 3 feet high and 10 to 15 feet across. I swear that he jumped it, and never touched the manure! Three or four of us saw it and next morning they measured the pile and it was just 2 inches short of 13 feet!

We got to the house, and down to the cellar just as a big one hit right near the base of the house. Several other guys were down there including Colonel Brownlow, his driver, Paul LeReux, (Campbell, NE) and Henry Duer (from Staplehurst). Immediately someone started down the stairs saying "Bitta, bitta," Please in German. The Colonel yelled, "Halt!" He sent Duer to talk to him in German and Le Reux in French. Neither could say a word. The Colonel said, "Damn it, Duer, say something or I'll kill him." Turned out it was the old man and old woman who lived in the house and they wanted in the cellar, too. If it had been a German soldier he could have gotten us all with one grenade.

There were some fresh garlic bulbs stored in the basement. Jackson ate some. You couldn't get near him for two days.

(Things like this, strafing, the guy jumping the manure pile, incoming shells, machine gun tracers, and so forth, all seemed to be happening in slow motion because you were so excited and your blood pressure was up so high everything you saw had a pinkish background. Like seeing red when you're real mad.)

The morning of November 22 we pulled through Saarburg, on to Reding. Saarburg was under heavy German artillery fire but no one was hurt. There were lots of arguments about who had the closest call. We moved out of Reding the same morning and kept changing positions as fast as we could. We were in position at Mittlebrown, Boursheid, and Eshbourg. In Eshbourg we grabbed a bite of turkey the cooks had been roasting on the mess truck. It was Thanksgiving Day. I'm sure everyone was thankful to be alive and not laying in the mud in some rain soaked field.

We then moved to Phalsweyer but this was bad. Somehow we had gotten behind the German lines. We made a fast return march clear back to the outskirts of Saarbourg!

The 44th Division was being counter attacked by the Germans from the flank and we were to help stop it. It took from November 22 to the 26th to get everything under control and to get the front line straightened out. There was heavy fighting and many small skirmishes, each side looking for a soft spot in the line. There were lots of small patrol on both Allied and German sides, lots of flares at night and lots of sporadic artillery fire on both sides.

When things quieted down, the Service Battery agent, Wagner, and I were out driving around one day, which we often did when we dared. We had his Jeep because Service Battery was usually back several miles and it was too far for him to walk to Headquarters. He would volunteer to take me back to A Battery when we got the messages done. He'd say, "Let's go look around a little." Several times we got off on a wrong road and got up with tanks and infantry, and once almost into the German lines.

This afternoon we came upon a small village that was completely leveled, every building burning. There wasn't anything over 3 or 4 feet high standing. There was dead livestock, dead civilians and soldiers. Not a thing moving.

We got to the other side of the village and decided we had gone too far and had better turn around. Then we saw some GIs about 200 yards down the road, tanks a little further on, so we drove on and talked to some of the infantry. They said they had flattened the village by tanks and artillery and had just chased the Germans out of there. The Germans were ahead down the road and patrols were out to make contact with them again.

We decided we had seen enough for this day. We finally found the road back to headquarters. This happened all within a small area, less than two miles wide and maybe only 3 or 4 miles deep.

Colonel Brownlow always stayed close to the infantry. Most of us thought too close in case the Germans broke through. When they did, we always managed to get out before they could overrun us.

The roads were crooked and there were many of them. It was easy to get mixed up in directions unless you constantly watched the compass. Many forks and tees in the roads. We were probably gone for a little over an hour. We would have been in big trouble if we had ever been caught on any of these sightseeing trips (of which there were many). Our pre-arranged story was always that we must have taken the wrong road - - - which was true many times, but not always.

The Service Battery agent was the only agent who had his own Jeep. A, B, and C Battery agents walked back and forth between the firing batteries because it was usually less than a mile. If farther, the Service Battery agent would take us in his Jeep.

Wagner was agent for Service Battery all through combat and I was agent for A Battery all through combat. Mauritz Johnson (Rockford, Illinois) had been agent for B Battery. When they took Jackson off the Forward Observer Team because he was a nervous wreck, had shrapnel in him, twitched and jerked when he slept, they put Johnson on the team in his place. Well, Johnson went up just once and came back so nerved up he refused to go again. They did not court martial him, but put him on permanent K.P.

Jackson really should have been pulled out of combat. He and I slept in the same pup tent when outside, and close by when we were in a building. He thought he was safer when he was with me or the C Battery agent Campbell, because we didn't get excited so easy, even though we got scared, too. Maybe he didn't know we got scared. He always wanted to stand guard with me. He stayed as B Battery agent during the rest of the war.

The first agent from C Battery was Leno Borino (Diamondville, Wyoming). The first two or three weeks of combat we pitched our pup tents together. He was soon replaced by Campbell, who had been with Forward Observers, and was wounded slightly (Bronze Star as well as Purple Heart.)

Borino was sent up with the Forward Observers. They got him in a tight spot, and he had to take over for the officer and direct the fire. A sniper was after them and Borino shot and killed him in an upstairs window. He said the sniper came falling out just like in the western movies. Borino got a Bronze star and a battlefield commission, as 2nd Lieutenant. He was the forward observer from then on.

C Battery sent another agent, Zazadil, to the Message Center, and he and Campbell rotated being agent and going up with the Forward Observer from C Battery. They were both kind of wild.

It seemed like all Zazadil could think of was something to eat or drink. He was always on the lookout for food or drink and writing home to have them send him food. He was always coming up with eggs or a chicken or potatoes or cabbage. You were not supposed to swipe anything from the natives but lots of guys did. (The 101st Airborne even butchered cows and pigs. They claimed the animals got killed out in no man's land but everyone knew better. They would bring meat back to the Artillery to cook it for them because they did not have a kitchen of their own on the line. Just had K rations to eat. So they would bring back a quarter of beef and share it with anyone who had a kitchen, usually artillery, and that battery would have a big feed.)

Zazadil knew about all the different wines and liqueurs. Mainly he liked cognac, burgundy, calvados and some of the fancy liqueurs and champagnes. When we would search houses he always seemed to come up with a few bottles. That night we would sit around a candle at a table and drink. If we had some cigars we would dip them in the wine and filter the smoke through the wine soaked end. Not bad.

But his main thing was food. He liked melba toast and finally started getting a couple packages of food from his family in Chicago every week. Most of the contents were melba toast. They must have created a shortage of melba toast in Chicago.

It was pretty good. So good that I had my Dad start sending some when he could find it. We had a joke about it when we got a package from home we would say, "Melba toast time."

(Overseas packages had to be a certain size. They were about 4 inches high, 8 inches wide and maybe 15 inches long.)

Command Car

Arthur Nickel, Chicago, Illinois was the driver at that time. I rode in the pile in the back seat! Captain Reginald Carpenter, South Bend, Indiana, rode next to the driver. It had a huge two-way radio that had short wave and could pick up stations in the United States, lots of music, big band music, and news from the States, the BBC and Edward R. Murrow.

CHAPTER TWELVE

In England, and for about three months in France, we got by sleeping with just our heavy wool blankets and a shelter half. (Half of a pup tent each man carried to be fastened to another man's half to make a pup tent.) Each man carried a half plus some light rope and half of the tent pegs needed to set it up. When the weather was good you didn't bother to set up the tent, just put the half on the ground and slept on it.

Now when the weather was getting colder in November, and wetter, the firing batteries got large pyramidal tents for each gun section to sleep in – a tent for 8 or 10 men dug into the ground three or four feet depending on how much time you had and how hard the ground was.

These big tents were set up over frozen ground, mud, or snow. Something had to be put on the ground or the bedding would get muddy and wet. Anything that could be found was used: wood, tin, bricks, straw, pine tree branches tarps, etc. The crates the ammo came in were used to kindle fires. Once the stoves were going in the tent, the ground would thaw out and get slick and slimy. You could hardly stand up in there. The guys on the gun crews had a terrible miserable time.

By winter I was staying with Headquarters Battery in houses or cellars. I was really lucky. Even if the houses weren't heated, at least they were dry.

In about November we were issued snow pac rubber boots and got out of the leather combat boots. The snow pacs kept our feet dry and cut down on the chance of trench foot.

We were also issued a pod type sleeping bag that was kind of body shaped and was fairly waterproof. It had an inner lining that was supposed to hold your body heat and keep you warm. It worked good but the bag was hard to get out of if we came under fire. As it got colder further into the winter this bag was not warm enough, so like many others, I cut up a wool blanket and sewed it inside the bag.

I also sewed a blanket in my combat jacket, sleeves and all. That really helped. It was easy to pick up extra blankets and other things that were left behind when someone was wounded or killed. Infantry threw blankets away because they didn't want to carry them.

We always slept in our clothes. You put the shelter half down, then a blanket folded to about four thicknesses, and placed the bedroll and a blanket over that, and then your heavy wool overcoat over that. You left on your head the knit cap that went under the helmet liner and helmet.

With all this I stayed pretty warm, but it was hard to get out of all that, pull on your boots and coat to run to a cellar or basement when there was sudden incoming artillery fire.

Fox holes and slit trenches were sometimes a work of art in places when there was time to build them. Normally the hole was dug just deep enough and big enough around to squat down in, and have your head below ground level with room to get your pack and gun in also. The slit trench was long and wide enough to lie down in and sleep. The danger of a slit trench was that a tank could run the length of it and crush the guy in the trench.

If there was time, the holes were covered with anything available, and dirt put on top of that. A few small pine tree branches might be put in the bottom so you didn't have to sit or lie on the damp earth. Some put heavy logs or boards over them, anything to keep out the shells and rain and snow. Some guys even made little ledges for cartridge clips, cigarettes and so forth.

When the ground was frozen and you couldn't dig, or if there was snow, you might get on the sheltered side of a tree away from the wind and pile snow in a box shape, put tree limbs on the bottom and sleep in this slot above ground with the snow built up around you to keep the wind off.

If we were stalled in one place for quite a while, some of these places were made pretty comfortable.

None of these survival things had been taught in basic training. It was just instinct. (Incidentally about the time we finished basic training in Camp White, the Army had decided after fighting in North Africa that 13 weeks of basic wasn't enough and changed it to 17 weeks.)

The 5 cooks in the battery's kitchen prepared our food. These guys had it pretty hard; no matter what was going on they had to come up with 3 meals a day.

There was always plenty of food, even though it might be K rations if we were moving real fast. (Crackers, canned scrambled eggs, hash or stew, a packet of coffee, packet of toilet paper, packet of cigarettes, maybe a small chocolate bar.) Everyone had a small key can opener to open the cans, called "the key to the ETO." Some cans had beans, some sardines. They contained various other items at times, like a tiny tube of toothpaste. The ration box was about 8 or 10 inches long and 4 or 5 inches wide.

There were always K ration boxes in the trucks and in the kitchen. The boxes were a wax-like cardboard, and items inside were wrapped in a heavy wax paper. We carried 2 or 3 in our musette bags.

The cooks prepared hot meals out of their mobile kitchens. Breakfast could be powdered scrambled eggs, hot cereal, and coffee. Other meals would be a variety of canned food, vegetables, beans, stews, canned hash, sometimes ham, maybe fruit. Bread came in at times from bakeries set up to the rear of the fighting front. The cooks had flour, sugar, etc, and sometimes made desserts like cobbler with canned fruit. There was always plenty of coffee.

Mail followed us real good. We had mail call every few days, sometimes every other day and even every day. I corresponded with quite a few people. Rosie Pokorney from Shelby who I had known when as a little girl she lived in Surprise. I liked her very much and I think she liked me but we never dated. Her letters were great. We wrote about every ten days or two weeks.

I had had dates with Mary Sievert (Shelby) before I went into the Service and we wrote back and forth for about six months or so while I was in the States, then tapered off. I may have been the one who quit writing.

Ahlene Debaetes, another Shelby girl I dated when home on furlough, exchanged letters with me from the later part of 1943 until past the end of the war late in '45.

Glen Olson, a neighbor I went to school with when growing up on the farm wrote a few times. He sent me a bunch of addresses of girls from Shelby, Surprise, Gresham, and other places and wanted me to write to them. I put the names in my address book, but never did write to them.

I would liked to have written to some of the nice girls I knew from country school, but most had moved away and I didn't know addresses. I can think now of about 25 girls near my age from our big country school whose parents had been forced to leave the area because of the drouth and Depression, and just as many boys.

Charley Eppenbaugh (Gresham, NE) and I wrote to each other. He was an MP in Hawaii. It took quite a while to get letters back and forth from Europe to Hawaii!

I exchanged letters with Lester Hans (Utica, Nebraska), Bud Hans' older brother. We went into the Army about the same time but he wound up in the Pacific Theater. It really took a long time for our letters to get back and forth. He was really sick of the war and afraid he would not make it back home.

Richard Block, a farm neighbor, was in North Africa, then Italy, in the Air Force as a tail gunner in a bomber that made the bombing runs over the Polesti Oil Fields, one of the worst runs in Europe. In one of his letters he had 23 of the required 25 missions (after which he would get leave to go home. He said he didn't think he would live through 2 more missions as they had lost many planes on each of his missions. The tail gunner was most apt to be killed.) I didn't hear from him for awhile and I thought maybe he had been shot down. Then a letter came and he was back in the States, had had a furlough, and was back at a base teaching new recruits to be tail gunners.

An older neighbor, Anton Coufal, wrote to me every month or so. (This is not the Anton Coufal who married our cousin, Mary Walsh.)

Puzz Menze wrote occasionally. Some of the girls from Surprise wrote to me, Doris Armagost, and Doris Schrader.

James Block and I exchanged letters when we were both in France and Germany. We tried to get together but even though we were only 50 or so miles apart we could not get away from our outfits during combat. James did get to see Jean Parks (farm neighbor) a couple of times. She was a Captain in the Army Nurse Corps. She and Jim had lunch together in Paris once. (Jean stayed in the Army after the war and retired as a Major.)

Of course the family was good about writing. Opal and Harry Pratt wrote regularly, mostly Harry wrote. Bob and Sylvia wrote and that was mostly Sylvia who did the writing. And of course Mom and Dad wrote every week. Dad did most of the writing even before Mom got sick. When I wrote to him from Europe I would give him clues to about where I was, but he never figured them out. I could almost say the towns when Captain Carlson, our battery commander censored the letters. We had always been good friends. But after he cracked up over killing so many German boys and left us, the new Captain blacked out the sentences that might be a clue as to where we were.

Uncle Todd Knouse wrote fairly often, Aunt Bess Huff wrote once in a while. Aunt Alma and Uncle Lee Walsh wrote now and then. I think I got a letter or two from Aunt Eunice (Knouse) Brown, Uncle Bob Knouse, and Dave Huff. There were probably others who wrote to me. Anyway I got quite a bit of mail.

When I was still in the States, I told the folks not to send me any packages because you could only have what the Army issued in your footlocker. Anything I might need in the line of toilet articles I could buy cheap in the Post Exchange. Overseas they did send packages of things like cookies, crackers, melba toast, cheese spread, jelly, etc., even a cake a couple of times. Any time someone got a package, the contents were passed around among the guys in your section and it was gone in about 2 minutes.

Camouflaged Mess Tent and Kitchen Truck

Unknown Soldiers, Combat 1944 – 45

Three hot meals a day unless on the move and then K rations. Staff sergeant, three tech sergeant assistants, three tech corporals and three privates.

CHAPTER THIRTEEN

We looked into a barn one day and noticed a pile of hay that had a strange shape. We dug into it and found a great big open touring car. It was up on blocks and the top was down. Some of the mechanics wanted to get it running so they put some one hundred octane gas in it, found a tire pump in the tool box on the back of the car, pumped up the tires and got it down off the blocks. Then the battery was down. It had an inline twelve cylinder engine in it and the speedometer when converted to miles showed it probably could do somewhere between a hundred thirty to a hundred fifty miles per hour. They primed the carburetor, hooked a steel cable from a weapons carrier to the car and gave it a pull. They only pulled it a couple hundred yards and it started. Smoked a little for a while and then ran so quiet you could hardly hear it. It had six speeds forward. We drove it around, taking about 8 or 9 guys for a ride at a time it was so big.

Then somebody had a German officer's cap and put it on and someone painted a Hitler mustache on him with a fountain pen and they drove around with him standing up on the passenger side doing Hitler salutes. Everyone really had a good time that day. They got that car up to over one hundred miles an hour once but were scared to go any faster. They said there was plenty of gas pedal left and the thing would probably have done 140 easy. I had a short ride in it when the guy was playing Hitler. Glad I wasn't in it when they got it up to a hundred.

We played around with it for a couple of days, and then we had to move so we just pushed it back in the barn. Now there was a souvenir – if you could have just gotten it home.

Another time in a shed we found a German military motorcycle complete with sidecar. It had a bunch of stuff piled over it but someone dug far enough and found it. It was okay, just out of gas. Everyone was driving it around or riding in the side car for a few days. When we had to move on, they set fire to it and destroyed it.

I found a bicycle with a motor on it along the banks of the Rhine River. It was a little larger than a normal bike and had a little gas in it. It started okay. We got more gas and rode it around quite a bit.

I was riding it on kind of a beach area near the river, going pretty fast when I hit a soft spot in the sand and went flying over the handlebars. Skinned up my arms and legs quite a bit. I just laid there and cussed. When I got up I saw that the front wheel was bent bad, so I pulled out my pistol and shot a hole in the gas tank. I tossed a match in the gas and let it burn up. I had to walk about a half a mile back. Told the ones that had been riding it what happened and that there would be no more rides. They had heard the shot and had seen the smoke.

They all thought it was pretty funny I had found the bike and I had been the one to destroy it. They said it was a good thing one of them hadn't wrecked it, that I would have been really teed off at them. For several days everyone was asking me if I had any more horses with broken legs that I had to shoot.

After burning the bike, I washed the cuts and scrapes on my arms and legs and put sulfa powder on them. I was sore and lame for a few days.

When we were firing a lot and moving fast we missed getting to go through a lot of places. Probably missed something big, but we gave it our best effort. It was penalty of court martial if you were caught looting, but we called it "liberating" and everyone, even officers, did it. After all, didn't someone once say, "To the victors belong the spoils?"

Now here's something strange. In a lot of houses we searched, we would find a used Kotex (sanitary pad) under the pillow of the bed. I run on to this probably 50 times or more and so did everyone else I talked to about this. Why would they sleep with this under the pillow?" We even asked Germans but never got a logical answer. It must have been some kind of superstition.

At one time we were holed up in a small farming village and had taken over a peasant's house for the Message Center. House and barn, all in one building, with manure pile at the front door. Though the place was small we didn't kick the people out. They could use all of the house but we would be there too. We did let them have the bedroom to sleep in. There was the mother, dad, and daughter Annie. They dressed in coarse woven clothes and wore wooden shoes, just like in a story book or movie.

We were receiving incoming German artillery fire occasionally but it didn't seem to bother the peasants as much as it did us. They didn't seem to be afraid of us either. Just went on taking care of the cows and having their meals.

This was the first time I saw how the peasants had their meals, just one dish in the middle of the table, like peas, or cabbage, or potatoes. They had a dark brown or what we called black bread and wine or milk to drink. There was real butter from the cows' cream on the bread.

We were there several days and it rained night and day, but the peasants clopped around in the rain doing their work, with a heavy coarse woven wool shawl over their shoulders and a leather cap on the head.

This was one of our first contacts with German civilians and we couldn't talk to them very well. They spoke both French and German but that didn't help us any.

Later, there were many times when someone in the Message Center would say something about Echenburg Annie, like "You know Echenburg Annie never won a beauty contest, probably never won a style show either, poor Annie"! Or, "Wonder what size those wooden shoes were? Do you suppose, maybe size 16 or even 18? Did you ever wonder what her underwear was like, probably made out of that wool stuff? Or maybe she didn't wear any, did you think of that?"

No! Not really!

One of the corporals said, "By God, I would like to live like that after the war. Just let the damn world go by, not worry if the Germans or French or Americans are in control. Just farm a little land, take care of the cows, eat black bread, and when you have to go to the bathroom, just go out the front door, shit on the manure pile and let this damn rain flush it away."

There was seven of us that stayed at Annie's for just 3 days and nights but we talked about it for months.

Once during a slow period we were staying in a house in a town and everyone was cutting up 105mm shell casings making ash trays and other brass gadgets. We soldered a 50 cal. machine gun bullet in the center of the base of the cut off 105 casing right where the firing cap was. The 105 case had been fired so it was safe, but the 50 cal. had to have the bullet part worked out of the case and the powder dumped out. The empty case was then put in a machine gun to fire just the firing cap and then drive the cap out of the cartridge. This would leave a hole and you could drop solder in it to make the 50 cal. stand up, then work the bullet back in the case so it looked real. In reality it had no powder in it and no firing cap. We used an armor piercing bullet, not a tracer because a spark or heat would cause a tracer to blow up or shoot off like a sky rocket.

Campbell got his 105 mm base made, removed the bullet from the 50 cal. cartridge and dumped the powder out of it. He either did not know, or forgot, to put it in the machine gun to fire the cap. So he takes a hammer and a punch and starts to punch it out. The little copper cap went off, blew out the bottom side of the cartridge right through his hand between the bones just back of the fingers and lodged just under the skin on top of his hand, bleeding like everything. (This little cap was a little cylinder about 1/8 inch long and about 1/8 inch in diameter.) We cut the skin, took out the cylinder, put sulfa on it and wrapped it up and told him to go to the doctor.

He said, “Hell, no. I’ll wind up in a hospital and then in a Repple Depple (Replacement Depot) and then they’ll put me in the Infantry. I’ll take my chance on infection first.” He kept the wound washed in G.I. soap, which was mostly lye and would kill any germ. He kept putting sulfa powder on it and in about a month it healed up real good. (Which goes to show what you can do if you have to.)

I got a chance once to go to a rest area in Luneville, France. I was there 2 days. The food was a little better than out in the field and there were hot showers. They had beer to drink.

I saw a movie and went to a French dance. The music was strange and the people were not too friendly. Some of the GIs tried to get the French girls to dance with them but got turned down so we left the dance and said to hell with the French. Someone said we should have let Hitler have France.

Some places across France the people were real friendly and thankful. In other places they were cold and nasty and tried to ignore the fact we were even around.

Wagner and I were driving around two or three miles back of the front late in November of '44 and drove into a village that the Germans were still dropping a shell into every now and then. This was still in France. We were surprised that a few small stores were open. One had a few French Christmas cards in the window.

We went in and bought some. We didn't know what they said – only word we recognized was "Noel".

The old woman who ran the shop didn't want our French invasion money so we gave her some candy and gum. Well, she wanted cigarettes. We told her we would give her a pack of cigarettes for a bottle of wine. ("Vin" in French.) She went to the back and came out with a bottle and we traded.

Then she wanted to trade vin for more cigarettes. We told her we didn't have any more, which was a lie.

So we left with our vin and the Christmas cards. We just got out of the village and the Germans shelled it again.

CHAPTER FOURTEEN

I "liberated" a very good 2.2 lens 35mm Lieca camera out of a house somewhere. I wanted a Luger (pistol). Sergeant Fletcher, one of the cooks who had a Luger, wanted a good camera real bad. At the time the camera was worth more than the gun so we traded and he gave me a couple of other things and a little money to get the camera. (I lucked out on this because the Luger is now worth much more than the camera.) I registered this pistol at the Seward County Court House when I got home – a requirement.

I had two other pistols but I thought we could bring only one home – this turned out not to be true—I could have brought all three because our bags were never searched.

One was an automatic German Mauser 32 cal., like some German officers carried. I found it in a German home under a bunch of stuff in a kitchen drawer. The other was an Italian Beretta 32 cal., like some Italian officers carried. I got this one out of a house, too, hidden behind some books in a bookcase. I gave these two guns to friends who didn't have a gun to take home. I think one went to Nebraska and the other went to Utah.

I found my high-powered field glasses, which are night glasses, in the bottom of a wastebasket covered with waste paper, in a German Naval officer's home. There were really fancy uniforms in the closets. I didn't know the rank and I didn't think to see if it was the uniform of a submarine officer or a surface war ship.

The house was really fancy. His wife was there and she had a fit over our ransacking the house. Also there were two slave labor French girls serving as maids. They wore maid uniforms. They showed us the fancy dishes and fine silver the family used, and then they showed us the cracked dishes and tin cups they themselves ate from, eating table scraps. We saw the tiny room they had to sleep in and their worn-out shoes and personal clothing.

I said, "You can't stay here now. Do you know how to get back to your home in France?" They said they did. I said, "Well, pack up what you need and can carry and start walking. You'll probably get rides along the way."

We waited until they changed into their old ragged clothes and made small packs of other belongings. We gave them enough K rations to last for a few days and took them down to the road.

Just as we got there, along came several supply trucks, empty, heading back toward France for more supplies. We flagged them down and asked how far back they were going. They said 40 miles. I said, "These girls need a ride to get back to France." They said, "Sure, jump in the back." We helped the girls into the back of the big 6x6 and told them to sit on the floor and hang on. (I believe every truck driver in the ETO was a maniac, only speed they knew was wide open.)

The driver and assistant driver were both so young it didn't look like they were old enough to shave. Most of the drivers of other trucks were young also. One of our guys said to me, "How much younger are they going to draft them?" I said, "That's about it, this war will be over before summer."

The others laughed and said, "You're crazier than hell." I said, "Keep your head down and don't get killed. By summer, I'll laugh at you."

I tucked my fancy new field glasses under my combat jacket and we walked the few hundred yards over to the 242nd headquarters. It had been a good day.

The large knife with the Nazi emblem among my souvenirs came from a house also. I didn't find a sheath for it.

The leather dispatch case came off a German prisoner. He was an agent or messenger. All of the small items I have came off of prisoners or were found in houses.

I found a brand new copy of "Mein Kampf" written in German of course, hidden under the lower feather bed in a house. It is still in the original shipping box. The book seems to have never been opened up.

The way we searched houses was like this: Two or three guys would take a house or store and go in and really go through everything. Sometimes the people were there but most of the time they had fled before we got there. Mostly we took military things, or Nazi stuff. Occasionally we would find a watch, jewelry, or money, but the people usually took that kind of stuff with them. Some of the armor and mechanized cavalry got into towns first and they usually blew up safes and banks taking any jewels, American or English money. They found lots of German money, but it was worthless so it was left.

A couple cavalry guys had a big box of German money in their Jeep and gave me a bunch of it. They said, "Pass it around for souvenirs."

I knew six or seven guys in the Sixth Cavalry and one day Wagner and I came up on them. They had just sneaked up on a farm house where a couple German Officers and six or seven enlisted men were inside cooking up a bunch of food. Some of it turned out to be American food. These Germans had not put out even one guard. The Cavalry guys burst in and killed them all. It was really a mess. The Cavalry took watches and anything else of value the dead Germans had. Wagner got white as a sheet. I thought he was going to be sick.

We saw and talked to these guys several times over a two or three week period; pretty tough guys. One of them had a German canteen bottle he was sending home as a souvenir. I said, "There's a lot of those around, what do you want that for?" He had baited me to ask. He said, "Here's why." He took off the lid and I saw it was full of jewelry.

We searched literally hundreds of houses. Mostly it was for wine or liquor, next for military guns or knives, then watches or jewelry or whatever we could put in our pockets.

One time we got in a cigar factory and took cigars by the thousands. Everyone had cigars for several weeks.

In some of the big wine places where there were big casks of wine, like 200 gallon barrels, the GIs would fill up five-gallon water cans. Some of these big casks would be broken and the wine ran out on the floor. Our group never broke the casks, armor or infantry guys usually did it. I suppose they had developed more hate for the Germans than we had. Lots of beer out of the beer gardens went into five-gallon cans also. What everyone looked for was real good liquor in bottles, champagne, burgundy, schnapps, calvados and sweet liqueurs.

Campbell and I were out on a road one day as guide and guards. Mainly we were there to inform anyone that came along looking for the 242nd where they were. We were wandering around the area and found a couple of infantry M1 Garand rifles along the road with quite a few full clips of cartridges for them. Evidently some infantry were killed or wounded here, and when they were picked up, no one picked up the rifles.

It was kind of mountainous area and we got to talking about shooting up hill and down hill. I knew the distances up and down a hill were deceiving to adjust the sights to, compared to distance on the level. We were picking out different trees and so forth up and down the hills to see if we could hit them the first time with the rifles.

After about 15 or 20 shots we got pretty good.

About that time a Jeep came roaring up and some captain wanted to know what we were shooting at and were there Germans in the trees. We said no, just target practice.

He was nice about it but said we should not be shooting just back of the front as it made the guys up front worry about gun shots behind them. We hadn't thought about that and didn't do it again. When we thought about it, it was a stupid thing to have done.

We kept the M1's for a couple of weeks and then turned them in. They were too heavy to carry around anyway. (Some of the artillerymen did carry these M1's instead of our little light carbine.)

Two men from Headquarters Battery were snooping around a railroad yard and decided to see what was in a tank car. They opened one of the valves and what came out smelled like alcohol. One of them said it was probably industrial alcohol and it was poisonous. The other one said, no it was drinking alcohol. He should have known drinking alcohol wouldn't be in a tank car, but he took a drink of it and said it was pretty good, and took another drink.

Almost instantly he went blind. The other fellow didn't know what to do with him so he took him to the medics. And they didn't know what to do with him, and they took him to the hospital.

We never knew what happened to him after that.

When the contents of the tank car were checked it was found to be rocket propellant fuel used in the German buzz bombs. Everyone called this fuel "buzz bomb juice."

After that incident, everyone knew not to drink out of a tank car.

Every night as soon as it was dark there were low flying German Reconnaissance planes flying over our side of the front taking pictures with infrared cameras. The camera was mounted in the belly of the plane and as the plane passed overhead you could see the red flashes from the lens as the shutter opened and closed.

These planes were about the size of a Piper Cub like we had but they could fly slower and the engine sounded different. They flew very slow about a hundred feet above the ground. The pilot carried small anti-personnel bombs that he could drop over the side by hand if he saw a good target. It was mostly for harassment but sometimes he would kill or wound a few. We were not to shoot at them but when first in combat we did shoot at one and shot him down. It was easy to do but it gave away your position (and also by chance it might be American.)

One evening Wagner and I were snooping around and got lost. We started down a long sloping hill into a valley. We both decided at the same time we didn't know the road and it looked like Germans about a mile ahead.

I said, "Turn around and get the hell out of here." He said, "I just thought of that before you said it." He turned the Jeep around and went back up the hill as fast as it would go.

It was getting dark fast and there was a village at the top of the hill all blacked out. As we came into the village, here came a "Bed Check Charlie" right over us with his camera going.

He started to circle and I thought, "We are going to get bombed!" About that time, Wagner said, "Look at that! Someone is signaling to him from that upstairs window. See that lamp and the window shade going up and down? He said, "Shall we stop and check it out?"

I said, "ARE YOU CRAZY? Get out of this village and head for the trees!"

Charlie was making another turn over us but he didn't drop a bomb. He must have been busy reading the message from the window and hadn't seen us. He made three more passes or circles toward the window.

About a mile further back we found where we had missed a turn to take us back to the 242nd. We never told anyone about this for several days. Didn't want it to get back to the Colonel that we had been driving around.

Bed Check Charley was over us every night just as soon as it got dark. It was this way all up and down the front. Germany must have had hundreds of these little planes. They also used them at sea, finding convoys. They tried it over England, but when most were shot down, it was discontinued.

Captain Salskov was our forward observer in our Piper Cub plane. He was a giant of a man, about six foot six, big boned and muscular. He was the Battalion Intelligence Officer and was normally in Headquarters Battery but he did fill in for other battery commanders when they were on leave. He was very quiet and never seemed to get excited. Some guys called him the big teddy bear.

We had just the one plane and one pilot. There was just room for two in the plane. Salskov was so tall he could hardly get in the seat. They would go up in the plane and go back and forth just above the Howitzers parallel to the front line. The pilot would go as slow as he dared without stalling the plane and Salskov would use field glasses, find targets to fire on and would radio the directions and corrections to the battalion.

Once in a while they would drift too far forward and the Germans would start shooting anti-aircraft guns at them. There would be big bursts of flak all around them and they would dive down low and head for the rear. They had a lot of close calls.

One day, a few of us saw a Piper Cub doing this, not ours, but the battalion next to us. An ME 109 German Messerschmidt fighter plane got after it. He took one pass and missed. He went back up and took another pass at the Piper Cub, machine guns blazing. The Cub dived and pulled out of the dive at less than a hundred feet. The ME 109 was going too fast and could not pull out. He went straight into the ground. There were some explosions and a big ball of fire. There was hardly anything left of the German plane or the pilot. By this time the German Air Force was pretty well beaten and this pilot probably had not had enough training.

Got word one day everyone had to take turns going back about three miles behind the front and see a dentist. The records showed we had not had any dental service in Camp White, Yakima, Sill, Howze or Camp Shanks except in an emergency. When my turn came, I had 3 cavities.

The dentist, a Lieutenant, drilled out the holes with a drill that he had to treadle like an old time sewing machine. No deadening, not even an aspirin! I said, "Who the hell came up with a machine like this?"

He said, "For a first rate torture machine, don't you think she's a beauty?"

After the holes were drilled, and it was a slow process, he took little pieces of lead like a BB and had me bite down on them hard and push them into the holes. And that was it.

He said he couldn't do anything for my front tooth that had one side busted off. (I had broken it hunting ducks with my Dad's double barrel shotgun. I always used Peters Hi Velocity shells in that gun. When I shot at the ducks I accidentally pulled both triggers. It kicked me so hard the left hammer came back on my mouth and broke off part of a tooth.)

I have asked different dentists I have gone to over the years if they had ever seen such a drill. Only one had, a young dentist in Wickenburg, AZ, he had seen one in a dental museum!

---Oh, yes - - - I got one of the ducks!

Serving the Piece

105 mm Howitzer in full recoil

Firing into Germany

Combat - Spring 1945

Camouflaged 105 mm Howitzer

Camille DeBruyne, Longmont, Colorado

Maxwell Stoner, Red Cloud, Nebraska

Unknown

Combat 1944 – 45

Camouflaged 105 mm Howitzer

Helmer Anderson, Two Harbours, MN

George VanZyl, Fingers, TN

Ammo box at lower left had been manufactured at Hughes Brothers, Seward, NE – small world!

Combat 1944 - 45

CHAPTER FIFTEEN

On November 26, we started moving after a big barrage. Slowly, we headed for the Vosges Mts. We would be in one position for several days before moving on. We were at Metting a couple of days and then at Ottwiller. There was a Schnapps distillery operating at full capacity here. I think every soldier for miles around carted off all they could carry. This kind of news got around fast and there were a lot of them drank too much.

After Ottwiller the pace picked up and the battles got hotter. We took positions at Weislinger, Volksberg, and Montbroun until at Echenberg the attack bogged down. We had reached the reverse side (west side) of the Maginot line and the Germans were determined to stop us here. They held the higher ground with good visibility, strong defenses of forts, pillboxes and tank traps.

Our artillery just could not knock those pillboxes out. Even planes could not get the Germans out. We had forward observers within one hundred yards of them directing the artillery fire, which was way too close. One tiny mistake in elevation of the guns, or a mistake of not putting enough powder bags in the shell, or even one defective powder bag would have wiped out the entire forward observer party.

One Distinguished Service Cross, and a few Silver Stars were awarded to Forward Observer men for this action. Finally the Germans pulled out of the Maginot back to the Siegfried line. They were attempting to straighten out their lines. This was about Dec. 16.

We moved north to Bliesbrucken, France toward the area where a great battle was shaping up. It became known as the Battle of the Bulge.

On December 21st A Battery fired our first round to fall on German soil. The battle occupied positions 500 yards from the German border. The forward observer party was in Germany, the only ones to cross the border at this time.

The Colonel wanted the big brass shell casing from the first round for a souvenir. He came to the Message Center and told me to get it and bring it to him.

I went out to the gun position and they gave me a casing. There were lots of casings around. I thought I might as well have the first one and give the Colonel another one. So I kept the casing they gave me, and picked up another one for the Colonel.

Sometime later when everyone was making ash trays out of these casings during a lull on the front, I told someone in A Battery I thought I might have the second one because I supposed the first one was already switched when I was given it.

He laughed and said, "Hell, you might have any number up to a hundred or so. The Captain of A Battery has No. 1, the Exec Officer has No. 2, non-coms and privates all grabbed one apiece. I would guess the one sent to the Colonel that you came out and got was at least 50 or 60!"

I guess the one I have might be one of the first hundred fired the first day into Germany. I do know it was actually fired into Germany that day.

We moved to Bliesbrucken to relieve part of the 3rd Army that had gone further north on the bottom side of the Battle of the Bulge. The lines were spread very thin because Patton had pulled many men and much equipment out of the battle to move north on the Bulge. Everything that was left was gradually moving north and the line was getting thinner and thinner.

It was December 23, 1944 by now and things got so hot we had to retreat back to near Achen for better defensive positions.

It was fairly quiet for Christmas dinner on the 25th. On the 27th we were relieved from the 44th Division and attached to the 106th Mechanized Cavalry leaving Achen. We had to move 35 miles under cover of darkness to occupy positions that had been selected for us. The Colonel was not too happy about this but it did work out all right.

We were doing all the firing for the 106th because they did not have any artillery of their own. We were in Germany but not yet across the Rhine. Hitler had taken this area from the French early in the years he was in power.

On New Year's Eve the Germans launched an offensive across the Saar River and we started firing about 2 hours before midnight. We had been sliding to the left along the Front and were set up in town. I was at Headquarters; the firing batteries were at the edge of town. There was almost a full moon, very bright outside that night and Bed Check Charlie was going back and forth taking pictures. We could see the infrared light of his camera as he took pictures of our area.

Word came in that an artillery outfit somewhere had fired some mustard gas shells on the Germans and they might retaliate, so everyone was to have gas masks ready. Well, there was hardly anyone who had a gas mask among us messengers. They were out in our truck at the firing battery, so we went through German houses and found civilian masks to have for the time being.

There was quite a bit of shelling going on and we could hear machine gun and rifle fire. Several hours passed; it was well past midnight. Radio Silence was broken by

"March Order, Baker, Charlie, and Abel." A battery, Abel, my battery was to be last to leave and would retreat with the mechanized cavalry that was doing the infantry job in front of us.

I was told to move with Headquarters and not to go out to Abel Battery. We pulled out in minutes. Telephone lines were left; the Germans had already cut them. They had broken through the Cavalry and were already in Charlie and Baker Battery areas. One of the batteries had set up their kitchen in a house and had New Year's Day's dinner, turkey and all the trimmings on a stove already cooking. The Germans came in the back door and the battery cooks went out the front. - And left all the food to the Germans!

Headquarters was on the road and Charlie and Baker right behind them. They went into position again about 2 or 3 miles back, then here came Abel and went into position. They were all much closer together than normal, this time for mutual protection. Fire power of twelve 105mm Howitzers, about fifteen 50 caliber machine guns, maybe two hundred seventy 30 cal. carbines and thirty to fifty 45 cal. pistols. These batteries are usually several hundred yards apart but now were almost in line, the twelve Howitzers in a slightly V shaped row to fire direct fire on an order of "Action Front." Or the order might be 6 guns fire left on "Action Left", or 6 to fire right on "Action Right", or to turn completely and fire all 12 on "Action Rear".

The 50 cal. machine guns were set up on the sides to fire left, right, forward, or rear.

Headquarters was about a mile farther back, no phone lines were put in and we were on radio silence again.

We messengers were ready to go find our batteries when radio silence was again broken. The Cavalry was falling back once more and so were the firing batteries.

There was not much firing being done by our batteries except on map coordinates, cross roads and towns. The forward observers could not see in the dark to fire on targets. "March Order" came and messengers again went with Headquarters.

We went back several miles this time. We went into position again and just set up guns, unloaded ammunition and put out guards.

This setup did not last either. I had started walking to where I thought Abel Battery and the rest would be with a message for the batteries. On the way I met a Jeep and it was my Captain from A Battery. He said "Don't go any further, jump in the back, we are retreating again and I'm going back to find a place to retreat to." I said, "You can drop me at Headquarters."

He said, "They are already moving back and we will be back further than even Service Battery, (usually 10 to 20 miles behind the guns). We may catch up with Service. If we do, you catch a ride with them, and if you see Headquarters, go with them. But if you don't see Headquarters, just stay with Service Battery and go right on back with them. You can't take a chance on getting left behind in this retreat. At the rate we're backing up, we will be clear back to Paris by tomorrow."

We caught up with Service Battery and I went with them clear back, about 40 or 50 miles, then laid down and slept the rest of the night (about 2 or 3 hours).

The firing batteries had kept backing up a few miles at a time until daylight, about 35 miles, headquarters about 37 or 38 miles.

The minute it got light enough to direct artillery fire on enemy targets, the enemy advance was halted, and the firing batteries dug in the guns and tents. They took turns eating and sleeping. The battery that lost their New Year's dinner had to eat C & K rations!

During the morning of January 1, we were strafed by three 109s. We hit one with 50 cal. machine gun fire. It caught on fire and crashed behind enemy lines.

By the night of January 1, we had contained the attack and driven it back a little in some places. Forward observer parties said they could see the whites of the German's eyes as they came across the bridge over the Saar River.

("Memoirs" 242nd Field Artillery Battalion by Ernest Eller, Coeur D'Alene, Idaho, Sergeant and message center clerk, says that only A and B batteries retreated but this is wrong. I remember the whole battalion retreated or moved laterally three times that night.)

The batteries fired out of their dug-in positions in the daytime as a battalion. Then at night one or two Howitzers from each battery with their crews were pulled out and sent up and down the roads parallel to the front, to the right and left of the dug-in position for a few miles. They would stop at crossroads and towns, bridges, set up and fire a few rounds haphazardly across the front into German territory. Just enough rounds for the German sound and flash outfits to pick up the sound and flash, and get it mapped. Then we would leave.

In a few minutes this spot would be fired on by the Germans but we would be gone to another spot. Now this was not done to get them to fire on a false target. It was done because we were the only artillery covering about 30 miles of the front with very little armor or infantry in front of us. The constant moving and firing made the Germans think we had lots of artillery.

Patton was sliding everything to the left to stop the Battle of the Bulge and rescue some units that were surrounded at Bastogne. We were part of the southern edge of the battle. (If we had gone a few miles farther north we would have gotten a battle star for that battle.)

This sliding to the left or the right is not a new idea. General Grant during the Civil War was a wizard at this. He was always moving to the left, leaving a thin defense line behind that stayed very active for its few numbers. Meantime the main force was sliding left, feeling for a likely spot to break through. The enemy had to slide to the right in front of the army that was sliding to the left.

We stayed in this area for some time. The firing batteries moved a little way a time or two to keep from being shelled or strafed and to keep away from photo planes.

Things were fairly quiet for a week or so and we became acquainted with the people of the town Headquarters was in. We used a Browning automatic rifle when on guard duty but everything stayed quiet. Just the worried officers coming out at night to see if everything was okay.

We had 2 or 3 officers who were scared most of the time. They did not even go very far from the building to go to the bathroom they were so scared all the time.

The biggest excitement on guard was hearing the townspeople come out of their houses and relieve themselves on the manure pile in front of the house.

While we were here, there were some of our P47s and P51s doing some strafing and dive-bombing ahead of us. German Messerschmidts got after them and in order to get more speed one of the planes aborted his two 500 lb bombs. They fell in our gun battery and both bombs went off. Luckily the soil here close to the Rhine River was mostly soft sand. The bombs' fuses were evidently set on delay, and they went into the soft sand

and blew a hole about 20 feet deep and about 60 feet wide. You could have put a big 6x6 truck and a couple of Jeeps in each hole and still have had room to spare. The sand almost covered one truck and a 105mm Howitzer and completely covered a dug-in 12 x 12 tent with the men in it. They had to be dug out with shovels. Unbelievably no one was killed and only a few had minor injuries. They couldn't hear very good for a few days.

If this had been hard or rocky soil, or if the fuse had been set for "quick", it would surely have killed 80 to 90 men around and knocked out 7 or 8 6x6 trucks, four 105 Howitzers, 4 or 5 Jeeps, a command car, and 3 or 4 weapons carriers.

Headquarters stayed in a nearby town and many of us at Headquarters got very sick with fever, vomiting and diarrhea. I was so sick and had runs so bad, and of course, no toilets or inside plumbing. We would run out the back door, drop our pants and go in the garden! It got so thick you could hardly find a place to squat. Lucky it was below freezing and it all froze.

There wasn't anyone who would go on sick call. You might get sent to a hospital and when you got well be sent to another outfit. You might get transferred to armor or infantry. I was so sick I went out of my head with fever and the other agents thought I should go on sick call but I refused.

The Service Battery agent said he would handle my messages to and from A Battery so I made it okay.

It was in this area that we had a white phosphorus shell blow up in the A battery gun site. It was below zero weather and very icy underfoot. PFC Juan Barrantoes (Bastrop, Texas) slipped and fell while loading a white phosphorus shell. The shell hit the frozen ground and went off. Barrantoes lost both of his legs, burned and blown off. He may have survived the burns not only because of the good care received from medics, but also because it was so cold. Several of the men were burned but not badly. The number one gunner PFC Harold Sharp was killed. He died in just a few minutes, before they could even get him in an ambulance. (I have since asked many of the A Battery men about him but none knew anything about him or when he came to A Battery or where his home was.) Barrantoes was conscious the whole time and said to his good friend Max Stoner (Red Cloud, NE), "Don't worry about me Max, it's white sheets for me in the hospital and home to my warm Texas."

Finally, Patton had moved to the left enough to free Bastogne. The front had quieted down except for sporadic artillery and infantry clashes, and a fire fight now and then caused by German and Allied patrols running into each other.

Intelligence and patrols figured out where the Germans were building up again for an attack and everything starting sliding to the right this time. We had to move about 100 miles to the right in the middle of the night. It was below zero and we ended up near Strasburg in the Saverne Gap.

I was riding in the back of the Battery Commander's Command car. I had on pure wool long johns and wool socks, also snow-pack boots, wool overcoat, wool pants and shirt, cap and helmet and the battle jacket that had the wool blanket sewn inside. I wrapped my legs in a blanket. When we got to our placement, I could not stand to get out of the car. Quite a few of us had the same problem; our feet were frost bitten. We finally got into a building and started the one-burner gas stoves we carried, got our boots off and the frostbite out. I had real sore feet for a couple of weeks.

The reason we had to get there so fast was because it was a hot spot and the Germans were building up in the area. The 101st Airborne had been brought in and they didn't have their own artillery. We had to get into position and fire for them.

There was fighting about every day and night (at night under flares from flare guns or veri-pistols.) After a few days it settled down to sporadic fights again. The Germans moved to their left deeper into the Vosges Mountains in front of the US 7th Army and the Free French and the American Japanese Battalion. Things really got hot to the right of us.

The Forward Observer party brought back a little boy German soldier they had captured. (Hitler was getting so short of men he was putting units of old men and little boys on the front line. "Volkstrum" or People's Army.) He was about 14 or 15; spoke perfect German, French and English. He had been educated in England and spoke English with an English accent.

A Battery decided to keep him as an interpreter. They moved him in with 1st Sergeant Reed and Herk (Helmer) Anderson (Two Harbours, MN). Someone even went and stole some civilian clothes to get him out of his German uniform so he wouldn't get shot accidentally. We had him for about 3 weeks and he seemed to like being with us.

But then the Colonel found out about it and really raised hell about breaking rules of the Geneva Convention by keeping a prisoner at the front in harm's way. He said he didn't want such a dumb thing on his records, if we got caught with the boy.

So Wagner and I went out to A Battery to pick him up. They had him put his German uniform back on and we took him to the rear. He said he really liked being with Sergeant Reed and asked what would happen to him.

We told him he would be sent as a POW to the United States and that he would be fine. He said his father had been killed on the Russian front and his mother in a bombing raid.

We had the best supplies, food and PX articles and movies every day while we were with the 101st Airborne. Also the best protection in front of us that we had all through the war, but after about 3 or 4 weeks they were pulled out and sent back for a rest before the Spring Offensive.

CHAPTER SIXTEEN

The 42nd Rainbow Division was put in front of us. We moved up fairly quickly close to the Rhine River.

We had moved a lot further north than we had been for some time, but not as far north as we were during the Bulge. In these moves we switched from one Division to another or to mechanized cavalry units and also switched Army Corps. Switched back and forth between Patton's Third Army and Patch's Seventh Army. The reason for the switches was that we were right on the line between these two armies and we went where we were needed most; usually a hot spot or to support a new division that came on the line whose own artillery had not arrived from the seaports yet. The infantry of these divisions were unloaded first and shipped non-stop by truck to the front. Some had not recovered from being seasick yet. A very sad situation, the last few miles they had to walk to the front lines to relieve the worn out men up there, these new guys would leave a trail of equipment. They threw away what they thought they were unable to carry, gas masks, clothes, even blankets and packs, and even ammunition. It would be several days before their own artillery would get unloaded, supplied and convoyed to the front from either the Mediterranean ports or the Channel ports.

Their artillery was as green as the new infantry, not as well trained as the 242nd and of course we had been in combat for several months by now. They would need to be under Corps artillery for awhile until they got some experience and could take over direct support of their own division infantry up front.

Sergeant Ernest Eller (Coeur D'Alene, Idaho) from the Message Center (the guy who wrote up all the recommendations for medals and who also wrote "Memoirs" 242nd Field Artillery Battalion) and I were standing in front of the Center and these new infantry men were walking by in single file at route step. They extended as far back as we could see and a couple miles on past the center. We commented on how young they were, mostly 18 and 19 year olds except the cadre of non-coms who looked older, but not much. The officers looked to be not over 22 and probably only a few months out of OCS (90 day wonders). We noticed them tossing away gear and that some looked sick.

Then we saw another long line coming from the other way from the front line, dirty unshaven and looking like they hadn't slept for days. These guys started drifting out of line, sitting down or even laying down. It seemed some were asleep almost immediately. There were kitchens set up so they could have a hot meal and coffee but very few went to eat.

Sergeant Eller said, "They are so young. How damn many years is this going to go on?"

I said, "Eller, you're the one with the high IQ, supposed to be so damn smart. Can't you look around and get the feel of things you see? This will be over in Europe in three or four months. Can't you see the massive buildup in this little area? Well, this is happening all up and down the line the whole length of the front. German air power is almost completely gone, except for an occasional strafing ME 109. We have hundreds of P47 and P51s bombing Germany every hour the weather is good enough. Hundreds of big American bombers are going over every day and hundreds of English bombers every night. They are not meeting the resistance we used to see and hear above us. Our small bombers are hitting German military just a few miles ahead of us, all up and down the line. And the Russians are moving in on them from the other side."

Eller said, "I think you are nuts but I sure hope you are right."

Activity was picking up on the German side of the front. The 42nd Rainbow Division and their artillery did not have any combat experience.

One afternoon it began to snow and by dark it was snowing hard, real big flakes. An hour or so after dark the Germans attacked and the 42nd Rainbow could not hold them. We got the word "March Order", move out and the order was C Battery, B Battery, Headquarters, and A Battery last out. It all happened so fast. The snow was coming down so heavy you could hardly see the narrow roads in the dark.

I went with Headquarters because I could not get back to A Battery before they had quit firing and pulled out. As we hit the turn at the edge of town to turn left to retreat, A Battery was almost at this cross roads and the engineers were blowing trees with explosives across the roads behind A Battery to slow up the Germans.

They continued to do this behind us for several miles through a wooded area. The infantry was already backing up right behind the engineers and armored cavalry.

We came to a Y in the road. Here stood two MPs and another sitting in a Jeep. They motioned for us to stop.

Our first thought was "Germans in American uniforms."

One of them said, "Give the password."

The officer in the Jeep I was in gave the password and asked for the counter sign. The MP gave it.

But this didn't prove anything.

Our officer said "You have a Texas 36th Division patch. Last I heard you were in Italy."

The MP said, "Just got in today. We're moving through to take over for the 42nd. They are in big trouble." Then he said, "You think we're Germans, don't you? How can I prove to you that we are the 36th Division?" Just then from the left road of the Y came a Jeep with driver, a Captain and a Major followed by truck after truck loaded with infantry all wearing 36th Division patches.

They really were moving through to take over the 42nd Rainbow. And just in time because the 42nd had really got beat up.

We didn't have to retreat any further.

I stayed with Headquarters and we went on about five miles to another town (Hagenau), took over some houses, set up the message center, fire control center and switchboard and then tried to get warm. The snow was getting worse and the wind was coming up.

Our firing batteries had set up and were digging into the frozen ground and firing back at where Germans might be by map coordinates, crossroads and so forth. There was no way to observe the fire or direct it in the snowstorm.

Some of the infantry in front of us began to drift back behind our artillery but tanks and rangers and mechanized cavalry were shifted over in front of us and helped hold the line.

We took one infantry man into our building. He did not even have a gun anymore. He said his company was either all killed or captured, probably about 80 to 100 men.

This was probably not accurate but they did get it bad clear up and down this area they were to control.

I believe there may have been a lot of panic. The whole division was in bad shape and had to be replaced. In

World War I this had been a very famous division (about 30,000 men).

By this time it was getting daylight, the snow had almost stopped but the wind was blowing up a blizzard.

Suddenly over the radio came "Enemy tanks and infantry in the open in the edge of town at Gross Helfendordf. Battalion adjust, shell high explosive, fuse quick, adjust center two of center battery, fire when ready." This voice was the forward observer with visual contact. (Breaking radio silence because no telephone lines up to the front yet.)

Immediately the radio operator at Center Two guns, Center Battery said, "On the way." A few seconds and the observer says, "Left so many degrees, up so much elevation," etc, and immediately "On the way" from the guns. Then another adjustment from the observer and "On the way" from the guns. A few more seconds, and the observer says, "Deflection correct, range correct, battalion fire 3 rounds for effect and immediately 3 voices say "On the way", and all 12 Howitzers fired 3 rounds or 36 high explosive shell.

Then from the forward observer, "Up – so many degrees, battalion fire 2 rounds for effect", and all 12 raise elevation and fire 2 more times or 24 shells.

This continued until the area was covered and German tanks and infantry were knocked out and the observer says, "Cease Fire. Mission accomplished."

Now the thing about this was: This was the town we had been staying two or three weeks in and we knew all the civilians there. Evidently the artillery fire drove the German military back into the woods beyond the town and the civilians started coming down the road in the blizzard toward our guns. The snow was about a foot deep with drifts 3 and 4 feet, and here they came walking and pulling carts with their belongings, old, young, middle aged and babies. Some were wounded and some killed and left behind.

When they reached our positions we put the old and young and wounded on trucks and took them back to the next town. Those who were able walked.

A little later the Germans came back into the town and we started shelling it again. This went on all day and into the night. Next morning was clear and bright. We brought in air support, bombed and strafed the town and shelled it again. Had problems here for several days. The village was almost leveled flat by this time. The Germans did not have a house standing to stay in. Perhaps a few cellars and the woods back of the town.

We fired for the 36th until their own artillery got there and then still supported them for several more days.

I believe it was around here that I first saw a jet airplane. The Germans had a few out in our area and the sound was really strange and the speed was terrific. However I believe these pilots lacked training on the planes because their strafing was not accurate and neither was their dive-bombing. They nearly always over-shot the target and did not get in close to do a good job strafing. I think they may have been afraid of the speed and did not come in close for fear of crashing. Also they had to go so far out to turn around for a second strike that everyone had taken cover before they came in for the second time. Most nerve-wracking was the first strike because at the speed of sound you didn't hear them coming. They couldn't do much in a dogfight with American or British fighters because of over shooting the slower fighters. The British Spitfire was faster than our fighters but they couldn't catch the German jet to shoot it down.

Anti-aircraft guns were not too good at getting them either. These jets were hard on our big bombers because they could hit the bigger target.

I saw a few dog fights or parts of them on clear days but they covered so much area and sometimes were so high that you did not see much of it.

On cloudy days we heard fights above the clouds and one day heard a real big one. I think we had big flights of bombers above us going into Germany that had a fighter escort. The German fighters must have come out to meet them. The noise started behind us on our side of the front and drifted on across the front on into Germany and faded out. At the speed they traveled it was probably a matter of minutes when a few parachutes came down in our area. One bomber crashed and a couple of fighters not too far away.

The bomber was scattered over a half a mile. The tail gunner was found dead still in his bubble. He had been hit by bullets from the fighters, the bubble and all around him was all shot up. No other bodies were found so they must have parachuted out.

Another time, moving up, we were convoying down a road when word came to hit the ditch, aircraft coming. We jumped out and scrambled to the ditches. It was one of our fighters, coming in, on fire, heading for a pasture parallel to the road about 200 yards out. He came in on his belly with canopy open. He had already jettisoned his belly gas tanks and the bombs under his wings. Before the plane stopped sliding, the pilot jumped out and took off running. The plane did not explode and he was okay when we picked him up. We continued to move on up the road. When we came to our designated position someone took him to an airfield about 50 miles back.

As A battery began digging in the Howitzers, I started walking to a nearby village where Headquarters would be. As I came up a little raise in the road, the German Artillery or a tank laid 3 rounds of 88mm artillery in around me. I heard the first one coming and jumped for the ditch. I had to crawl and run about a half a mile in the ditch to the town. While I was in the town they shelled it off and on all day and all night. There were German patrols in the town and we had a bad night on guard, did not get much sleep all night.

Next morning, more shelling and pretty accurate, too. Someone finally figured out that there was a German observer in the church steeple directing the artillery fire. Infantry got him and the firing stopped. They found the old preacher tied up and locked in a closet. The guard duty that night was about as scary as any all through the war.

After the Battle of the Bulge, the call came to cannibalize the outfits. (That is, take men from the units and transfer them to the infantry or armored divisions.) We had several guys to be transferred from A Battery. The morning they were to leave (about 10% of the Battalion) one guy from A Battery shot himself in the foot so he wouldn't have to go. They had to pick another person. He went to the armored division and lasted about a week. The tank he was in was hit and the whole crew burned to death.

Most of that 10% were either killed or wounded. We believe that only a few made it through without something happening to them. We lost contact with them after a while.

The guy that shot his foot came back to the hospital to headquarters. Wagner, the Service Battery agent and I were told to load up his gear and take him back to A Battery. So we did. Evidently they had called the battery and told them he was coming because when we drove into A Battery firing position, here comes First Sergeant Reed and Captain Carpenter, the battery commander, and the Exec. Officer Lt. Maddox.

Sergeant Reed called, "Kid, come over here, we got to talk." They were about 30 yards away. He told Wagner and this guy we were bringing back to wait in the Jeep. I got over to the officers and we walked about another 50 yards away.

Sergeant Reed said, "We won't have this guy back here anymore. Someone will kill him. It might even be me. The Captain and Lieutenant don't think there will be any trouble but I know there will be. So we have a paper for you to give to Colonel Brownlow to place him somewhere else. Take it to the Colonel and see what happens."

We drove back to headquarters. I gave the paper to the Colonel. He read it and said, "I'll let you know in a few minutes."

He talked to Captain Salskov and Captain Donald Hinman (Redlands, California). They decided to transfer him from A Battery to Service Battery, which would be back about 10 or 20 miles. So Wagner took him back to Service with transfer papers, which were typed up in a few minutes. We never heard of any problems with him there but he was never near any of the A Battery guys again.

History books say there were many SIWs (Self Inflicted Wounds) and General Patton's book says there were many. But this is the only one in the 242nd that I know of.

CHAPTER SEVENTEEN

February 1945

Dad wrote that he and Uncle Lee Walsh were going to sell their farms. He wanted to know if I thought it was all right. Although they both had mortgages on the land, they could sell and come out with a little money. Dad would make a down payment on a house in Seward and he and my brother Bob and his family would live there. Dad planned to go to work for CA Swanson Co, a processor of chickens, ducks, geese, turkeys and eggs in Seward. (In the building located near the railroad tracks on south highway 15 in Seward, now occupied by Graphic Printing.)

I wrote back that it was okay with me. I really didn't like farming very much anyway. The world had changed and things wouldn't be like it used to be on the farm. I had some thoughts about staying in the Service then and if I could have stayed with guys I knew through the war I might have stayed in but all were getting out.

I was walking from headquarters in a town out to A Battery about two miles in the country and was about halfway there. It was bright, bright sunshine and lots of snow on the ground. It was so bright it was hard on the eyes.

Suddenly two German jet planes began strafing an area about two miles away. There was no one else on the road so I got off the road, laid down and covered myself with snow in case the planes flew over the road. I waited until the strafing, anti-aircraft shooting was over and the sound of the planes died away before I got up and continued on to the battery. There had been terrible noise over there where the strafing was, but I never did hear how many or if anyone was killed or wounded.

It was difficult to bathe and wash living out in the open. GI's were expected to shave every day when at all possible. Most of the time we washed our face and hands and shaved in cold water. We each had one bath towel, one face towel and one wash cloth. It was hard to get them washed out and dry in combat so they got put back in your musette bag damp. They would mildew and get pretty bad sometimes. Of course there were times I didn't get to wash or shave for several days but General Patton was very strict on dress code, shaving and haircuts.

We wore wool dress uniforms but no tie or dress coat (called "blouse".) The wool coat we wore was called a battle jacket. Patton claimed you had to look good to keep up spirit and that teamwork in battle was better. He was probably right. At least you probably felt better cleaned up. And if you got killed you might as well be looking like a soldier instead of a dirty bum.

We took baths out of the helmet with just a wash cloth, and once in a while you could find a way to heat the water.

Wagner and I were back behind the Front a few miles one day and over by an ice covered pond we seen a tent, a couple trucks and trailers by it. There was steam coming out of the tent. So we drove over to take a look.

The Engineers had set up a machine that pumped water out of the pond to a big gas heated tank, then put the water under pressure and run into a tent. There was about six or eight showerheads.

The sergeant said, "When was the last time you had a hot shower?" We said "At least six months ago!" He said, "Help yourself!"

So we got our soap and moldy towels out of our musette bags and took a hot shower. It was great! But when we got out from under the warm water we damn near froze to death! The temperature was well below freezing outside.

We came out of the tent and asked the Sergeant in charge who these showers were for. He said, "Anyone. We just set up and if you want to tell your outfit, go ahead and tell them to come over."

We went back to Headquarters and told the Colonel. He was really tickled and started sending one truckload of men from each Battery over there at a time. By the end of the day everyone in the battalion had a hot shower!

We never saw one of those setups again. After the war, we heated water and used bathtubs in houses we took over. I have a picture of Corporal Rudolph Oeltjen (Fairbury, NE) carrying firewood to heat water in Assling after the war.

When we got a chance to wash towels and underwear we hung them up in the rooms we might be staying in or in the tents the firing batteries lived in.

Our wool shirts and pants we washed in 100-octane gasoline and hung them outside. This was just as dangerous as putting a wick in a gas filled bottle for a lamp, but I never heard of anyone getting burned.

I never did see anyone taking a bath or swimming in a river like you see in movies. I did see GI's washing their faces and shaving in street fountains in towns where the fountains still worked. We had to keep changing socks and washing our feet to keep from getting a condition called "Trench Foot". We also needed to dry out the inside of our boots by a fire whenever we could.

I picked up morning reports from Sergeant Reed every morning and delivered them to headquarters. Morning reports were how many men were present for duty, how many on sick call, how many in hospital, the inventory of gasoline, 105mm shells, food, etc. and the location of the Howitzers.

He asked me one morning how I liked my agent's job and if I regretted not going for sergeant's stripes in the kitchen. I said I sure liked the messenger work and was glad I had turned down the kitchen.

He would always say, "Good morning, Kid," or "How are you, Kid, what are the rumors today?" I was about the only source of local news for the guys in the battery and sometimes the only source for world news. I usually knew what Big Brass was in the area and that they might be stopping in. Many times I knew when we might be moving up or retreating so they would have some of their things loaded to get out in a hurry.

I was sure in February that the war in Europe would be over by April or May. (It ended in May.) They REALLY called me "Rumors" on that.

I was the first to tell them a plane that had strafed them was a "jet" powered plane, among the first in the world.

I was the one that brought the news that 2, 3, or 5 could go back to a rest area for a few days, or that 15 to 20 could take a truck back to a certain place for a movie that night.

When things were tight and there was a possibility phone wires were tapped, I would leave headquarters an hour before dawn and walk to "A" Battery with firing orders for a dawn barrage. This was very hairy in the dark. You better know the password when you get to the perimeter of the Battery, hope that no one is trigger happy, also that you don't run into another outfit, or maybe even a German patrol in which case the firing orders were to be destroyed.

I would get to "A" Battery and they would wake everyone up and get ready to fire on pre-arranged targets in the firing orders. The orders gave elevation, deflection, powder charge, etc. and how many rounds on each target. At the very moment given in the orders the whole front lit up like a terrible thunder storm. The smoke hangs heavy on the damp morning air. Some of the guns fire star shells and flares that come down by parachute and light up the main line where the German infantry is. After so many minutes of firing, our infantry would come out of their foxholes and start advancing and the artillery would slowly raise the elevation into a rolling barrage.

At times the gun crew would get calls from infantry radio or telephone from our forward observer who was with the infantry to zero in on targets they needed eliminated. In that case the forward observer directed the fire from up front, so many degrees up or down until the shells bracketed the target, hit it direct and knocked it out.

February 5, we moved laterally to Batsendorf, France and started supporting the entire 101st Airborne on their portion of the front. We could relax a little with them in front of us and not worry about the Germans breaking through them. They were pretty mean, especially the ones who had lived through D Day and the Battle of the Bulge. There were quite a few nasty firefights, especially at night.

March 2, we were relieved from the 36th Division and attached to the 103rd Division. We moved to Ingwiller, France, went into position and started firing defensive fire for them. This continued until the 15th of March with sporadic firing. Most of this combat seemed to be fire fights between combat patrols in efforts to straighten out lines for future attack, to find better jumping off places so they could move fast when the attack started.

On the 15th of March, the 103rd launched an attack along with a coordinated attack all up and down the front. Since March 2nd there had been a big buildup of supplies, ammo, etc. There were thousands of infantry, hundreds of tanks, hundreds of artillery pieces, tank destroyers, etc. There was a terrible artillery and tank barrage at dawn and by sunrise the infantry jumped off and started forward. This attack was meant to destroy or force the retreat of all the German army west of the Rhine River,

The battery agents went to the firing batteries before dawn with the orders chosen the past few days. As soon as the barrage ended, the infantry, tanks, and mechanized cavalry started to move and all our forward observers moved with them. The forward observers took over the firing for any target they saw, or the infantry needed taken out. There had been 1260 artillery pieces used in the pre-dawn barrage.

The first day we moved up to keep in range. We received hostile artillery fire on B and C Batteries near Mirhlhausen, France. (Sometimes called "Counter Battery") We moved rapidly on a push that carried us out of the Vosges Mountains and into the Siegfried Line. We were in positions at Offwiller, Niederbronn, Limbach, Climbach and Weiler all in France. By the 23rd of March we were again in Germany near Birkenhort. Then to Oppenhoffen, Germany and on the 26th of March we were relieved from the 103rd Division and attached to the 71st Division. During this time we fired over 3000 rounds from Weiler alone, some on the Siegfried Line, but mostly on the retreating Germans jammed up on the roads, mountain passes and narrow bridges. There was terrible destruction of men and equipment along these roads and many prisoners were taken. It was mainly done by artillery, as the weather wasn't the best for air support, maybe only an hour or so every few days.

We occupied positions near Limbergerhof, Germany on March 26th and fired across the Rhine River into

Mannheim and the area just south of the city. We shelled Mannheim for 4 or 5 days, plus P47s were dive bombing and strafing. There were a couple of days of heavy bombing by B24s and B26s, in flights of 12 to 24 planes in V formation. There was no air resistance by the Germans, but a lot of anti-aircraft fire came from them. This bombing was low level, probably not over 2000 feet or maybe less. I did not see any of our planes damaged from the German anti-aircraft fire.

Mannheim was shelled and bombed to ruins.

About this time Talavera got word that his twin brother had been killed in action with the infantry of the 79th Division.

Mannheim, Germany

Adolph Hitler poster on street sign

Combat 1945

242nd Field Artillery Battalion

Crossing the Rhine into Mannheim, Germany

The Germans had retreated and the city was in ruins

March 31, 1945

CHAPTER EIGHTEEN

March 31, 1945

On a warm spring day we crossed the Rhine River on a pontoon bridge at Mannheim, Germany. We moved down the highway in a well-spaced convoy. (Spaced in case of air attack, strafing or dive-bombing.) About 75 to 100 trucks, Jeeps, weapons carriers, command cars and twelve 105mm Howitzers. Mannheim was still burning as we entered the town. Did not see hardly any civilians. Guess they were still hid underground or had fled with the German soldiers.

We were headed toward Heidelberg, a world famous university city that had been declared an open city. It had not been bombed or strafed and there had never been any artillery bombardment or street fighting in this picturesque city.

The trees were just beginning to leaf out. We drove right down through the main part of the city. Wide cobble stone streets. Not a person or vehicle in sight. Rather strange, for being in an open city. The locals didn't have anything to fear from us.

We kept on going and as we got to the edge of the city on the other side, everything broke loose, in-coming mortar, artillery fire, but mostly machine guns. Tracer bullets were flying everywhere. Tracers really look strange in bright sunlight.

We jumped off the trucks and took cover along curbs, behind trees, etc. The trucks and other vehicles turned down side streets.

The mortar and artillery stopped just as fast as it started. There was some mechanized cavalry with us so they started to check out the machine gun fire. They moved up slowly but found nothing except empty shell casings. It had been a German rear guard. They had delayed us for over an hour.

We got lined up and continued on the road for quite a few miles - - until we ran into resistance and had to go into position and start firing.

We occupied positions at Fredrickfeld and later in the day at Plankstadt.

On April 1, we moved to Sandhausen where five ME 109s and one jet plane got into our area strafing. We had so much anti-aircraft, and artillery plus the hundreds of 50 cal. machine guns and rifles by the thousands that we drove them off. I don't believe there was any serious damage or casualties. At least neither I nor the other agents heard of any. We got around over the area quite a bit every day. Kind of had our finger on the pulse you might say.

On April 2 we moved to Weislock and did some firing, on to Hoffenheim, more firing.

April 3 we moved to Stienfurt and did quite a bit of firing to break up a counter attack, had it halted by morning and the Germans retreated. In fact we couldn't find them. This had probably been a rear guard attack instead of a counter attack. Anyway we were right with the infantry. Way too close to the Germans. We decided to haul all the infantry we could with us in our trucks and try to catch up with the Germans to make contact.

This was either daring, or stupid, or both. But this was typical of Colonel Brownlow. Always getting up a few hundred yards behind the infantry instead of a safer few thousand yards. After all these 105mm cannon could shoot several miles with great accuracy.

Anyway we lucked out. The Germans did not expect us to catch up so soon and we ran into small arms fire. Tracers were flying everywhere. It was still German rear guard. The infantry jumped off our trucks and started crawling forward. We drove off into some gullies and started firing just a few hundred yards ahead of the infantry that was only about 400 yards ahead of us.

We were in the edge of the town of Kirchhausen. The Germans backed through the town and crossed the Neckar River to the city of Hielbronn where they had strong fortified positions. We stayed here and fired over 6000 rounds into Hielbronn. Sometimes smoke shells to set up smoke screen for infantry and tanks to get across the river and to keep the Germans from blowing the bridges. There was artillery everywhere, firing across the river. There were battalions of 105mm and 155mm right next to us and long tom 155mm behind us firing over our heads. Infantry mortars close to the river firing across, and P47 planes bomb and strafing.

By the 12th of April things had calmed down pretty much and by the 14th we crossed the Neckar under cover of a smoke screen because there were still German snipers as part of a rear guard.

We were moving pretty fast, sometimes three or four positions a day. On a bright, warm sunshine filled day we were going down a road exactly like the one going east out of Milford, Nebraska. (You can drive out of Milford to the east along a road that goes along an embankment, turn right, go through a tunnel in the embankment, under the railroad.) We made a right turn and went under the railroad embankment above us. There several hundred yards ahead was a small village. To the right of the village was a big orchard, the trees all neatly spaced and in full bloom, pink blossoms. We jumped off the trucks and started checking buildings for a place to stay, or maybe a bakery, or a beer garden. A couple of guys and I ran into a place and grabbed up several bottles of booze. As we came out of the door, all hell broke loose.

German machine guns dug in place in the orchard were firing at us. Again, tracers were flying. There was wild confusion in the village street as trucks tried turning around to retreat back under the railroad and behind the embankment. Some smashed into the sides of buildings. Some guys didn't wait for the trucks and ran back the several hundred yards to the shelter of the embankment.

We started shelling the orchard, just popping the shells up and over the embankment into the village and orchard.

About that time a couple of American tanks with infantry riding on them came up behind us. The tank commander yelled, "What the hell you doing up here? There hasn't been any armor or infantry through here yet."

Some GI in the background said, "I wonder what the hell else he thinks is new?"

The infantry checked out the village. We listened and nothing was happening. Pretty soon a Jeep came back. The officers said they had found the places the Germans were firing from and they were gone.

We loaded up and followed them until they made contact with the Germans.

This was another case of the Colonel getting us too far forward with the artillery. We were lucky again as we had been several times before. A few skinned knees and elbows from hitting the dirt when the firing began. Some of the equipment had bullet holes but no one was hit. Didn't seem possible the Germans could have missed hitting someone what with 90 men milling around in all that confusion. I don't believe any of our machine gunners got a shot back at the Germans.

Later when we talked about it, we thought most of the bullets went over us. We wondered if the Germans may have been green or maybe they were Volkstrum Soldiers, little boys and old men, and they couldn't hold the machine guns level. A machine gun will "climb" if not sandbagged, or doesn't have a strong experienced man firing it. The first 3 or 4 shots will be straight and level, then higher and higher.

I will never forget that orchard.

We were on the front line for two hundred thirty odd days in a row and we worried about the older men and helped relieve them of some of their duty. Two of them made it through by sheer will power or guts.

Joe Freas was one of them he was a Portuguese from California, real short and rotund, not fat, but hairy? He had hair all over his body about 2 inches long and black as coal. His English was poor as was his family background. He had worked in the vineyards of California and had never seen snow until we were in Yakima and he could not believe the snow and the cold. He was in my tent in Yakima.

He was a real patriot and a real American. He would sit around and smile and play his accordion. He was so hairy he almost had to shampoo all over when he showered. He was constantly kidded and called ape, monkey, etc.

During combat I said to him one winter day, "How you doin', Joe?" and he said, "Well, Walsh, I guess I'm makin' it all right but the world is all wrong. I'm so far from warm California and home and I think I know why, but I'm not really sure. And I'm not sure I'll make it back home." He went on, "A night or so ago I came in from guard post to wake up the two guys to go on guard next while my partner stayed at the post. There was someone standing off to the side of their tent. I said 'It's really cold tonight, ain't it?' and he just grunted back at me. I thought it was one of the guys that had got up to take a leak but when I got in the tent, they were all there in the tent. I got scared as hell and woke up the one to go on Guard. We went outside and the guy was gone. The next morning we followed the tracks in the snow toward the German lines. About a hundred yards out they joined tracks of at least 3 others and they were German boot tracks. A German patrol had got past our infantry and was almost in our artillery position. If I had pushed the conversation with him further he would probably have killed me with a knife before I even knew he was a German. Now, I'm scared when I'm on guard duty because even though I can handle the 50 cal. machine gun at the guard post and also my carbine, I don't know if I could shoot them if I saw them."

I told him that if he seen any patrols again and if they were not too close that he should not get trigger happy because they were not looking for trouble but just out checking how many of us are over here in this side of the front. I told him I was pretty sure the war would be over in another 2, 3 months and the front was stabilized and would probably stay that way until spring, then it would not take too long to finish the Germans and the armored divisions and mechanized would do most of it.

It turned out I was right. The artillery could not keep up with the tanks the last month of the war but we were not very far behind and we did lots of firing on small pockets of resistance or groups that were bypassed by the armor. We were on the go night and day, going into position, firing until they surrendered, then getting a move order and moving on. Many times not long enough to dig in.

Anyway, Joe made it okay and got back to California.

Another older fellow was a radio telephone man, Hellmer "Herk" Anderson, and he made it also.

One day I was sent back of the front a few miles to a Division Headquarters Message Center to deliver a message and to pick up one in return. Field telephone wires had been cut and were believed to be tapped. As I walked along looking for the message center, a voice called out "Walsh, how are you and the 242nd doing?"

There standing on the steps of the house Division Headquarters was Major Goodrich (from basic training in Camp White back in 1942.) He was a Brigadier General now. Imagine – the head of the Division Artillery and he wanted to know about the 242nd and everybody in it. I could not believe how he could remember all the names and also how he could recognize me! Of course, I'm sure he knew the 242nd was in the area and that a message was coming by messenger, but he could not have known it would be Corporal Walsh.

We took over a four room house for our Message Center during this time and let the family, a man and his wife and 2 teenage daughters, stay in 3 rooms. There was a hillbilly kid in the 242nd. In fact there were several, but this one had "liberated" a guitar somewhere and he played it – really bad. Everyone got a kick out of it so one evening we invited him to the Message Center to play and sing.

He did a few songs and the German family just watched him kind of sour-like, but once in a while the girls would smile. And then he sang, "You Are My Sunshine", and they started singing, "Du Bist Mynon Sunenshon."

When he finished, Corporal Emerton told him to just play on the guitar, and Emerton kept motioning to the girls to get them to sing. Finally they did and then he couldn't get them to stop, just kept singing it over and over. They sure were better singers than the hillbilly kid.

(We had a cousin of country singer Roy Acuff's in the 242nd. He couldn't sing very good either!)

1st Lt. Edward Lazanby, (Baltimore, Maryland) and his forward observer group (driver, radio operator) were just outside of a town where there was a flock of tame ducks, 25 or 30 on a pond. There were also some infantry and armor guys there. Lazanby and his guys decided those ducks would make some fine eating. They couldn't get out there and catch them and they couldn't shoot them because the Germans would snipe at them if they got out in the open.

Lazanby said he could kill the ducks and keep the Germans under cover if some of the people would go pick up the ducks. Several volunteered but wanted to know how he was going to do this. He said, "Artillery, of course!"

They said, "You're kidding!" He says, "Just keep your heads down and wait until I tell you to go. "He went to the radio and said, "Fire Mission. Enemy infantry in the open. Battalion adjust. Time fire," and he gave the rest of the directions for the guns. A voice came back, "On the way."

Then there were twelve perfect air bursts twenty five to fifty feet above the pond. It got all the ducks!

He kept having the firing sections raise the elevation a little and walked the bursts right on and through the edge of town into the Germans. The volunteers ran out and got the ducks.

Lazanby not only got the ducks but the barrage was so severe the Germans packed up and retreated out the other side of the town.

One time Lazanby and his group was watching the German lines at quite a distance with field glasses. They noticed that about every hour or so two Germans on a motorcycle with a sidecar would leave a village and go to a wooded area about a mile and a half away and then return to the village. This went on all morning. He figured there must be German artillery or maybe tanks in the wooded area and probably a small headquarters in the village.

Never having any doubt that he was right, he called back for more battalions of Corps Artillery. The Colonel called in three more battalions of 155mm long toms and Howitzers. Lazanby gave them coordinates of the woods first and called for three rounds of white phosphorus from 48 of the guns. When all the shells hit, the woods were set on fire and burned up a couple tanks, a self propelled gun, some artillery pieces and other equipment. How many men we don't know. Those who were left got out fast.

He then had the guns turned on the village and flattened it with delayed fuses on high explosive shells, probably at least 284 shells, a very heavy concentration. Nobody seemed to leave the village but it was almost dark so some probably got away after dark. A harassing fire was kept up on the village all night.

We didn't move up for a couple days and then it was at a slant to the right and we never saw the area to find out just how bad the damage was.

Some of the 242nd men and infantry and armor men said Lazanby was the calmest they had ever seen all through the war. Infantrymen said they would rather be in the infantry than be in an artillery forward observer party.

We had been moving fast for some time when we pulled into a town and were told we could have at least a week's rest. The firing battery people were going to stay in a building for the first time since Camp Shanks, New York. They had been outside in tents all that time.

(Headquarters always took over houses to stay in, and I stayed with them most of the time, lucky me.) So every group kicked people out of their houses and unloaded their gear into houses, taking baths, washing clothes and really happy with the prospect of sleeping in a house. This was about 9:30 in the morning on a beautiful sunny day.

The whole battalion was in the town. Even Service Battery had come up with us. Everyone was visiting with friends in other batteries.

But it was too good to be true.

About four in the afternoon, the officers and non-coms were going door-to-door yelling, "March order, March order, Abel, Baker Charley and Headquarters. Service will stay in town. Saddle up, saddle up!"

It had been a very short rest. seven or eight hours at most.

It seems Patton and the Third Army were veering off to the northeast and moving so fast they didn't need artillery. We were changing army, changing corps, changing support to a different division that was in the Seventh Army under General Patch. We would head southeast.

Those men in the firing batteries never got to sleep in a house until a week or so after the war was over.

We were on the road almost night and day. We had been transferred back to the 103rd Division from the 100th and were rushing up and down roads firing in many different locations.

We came upon the concentration camp at Dachau, just shortly after the prisoners had been freed. The people were nothing but bones. The place smelled terrible. There were corpses piled up like cordwood. Prisoners had been dying faster than they could be buried or cremated. We were there only a few minutes and had to move on. Those of us that did see some of the inside of the place wished we had not and we were glad to move on.

(In recent years, 1990's, I have seen documentaries on television of three different outfits that claim they liberated Dachau, one Japanese-American regiment, one black armored battalion and one white regiment. Now we stopped there about 15 minutes after the camp was liberated and I don't remember seeing a black or a Japanese face there, only some white armored and infantrymen.)

CHAPTER NINETEEN

We started our final drive of the war into Bavaria. We occupied positions rapidly, sometimes two or three a day. At Weinsburg, Affaltack, Unterhemsback, and Ammetswieler.

We were rushing down a road with the Colonel's Jeep in the lead when it hit a land mine. Blew the whole front end off and clear off the side of the road, lifting it several feet in the air, pieces flying everywhere. Colonel Brownlow and his driver were thrown off to the side quite a few feet. Luckily they were not seriously injured, small cuts and bruised up pretty good. It had been hard on their ears and they couldn't hear very good for a few days. I was about a half mile behind them and heard the explosion and saw the dust and smoke. The Jeep looked like a train had hit it.

On the 19th of April we occupied positions in Hoppenbach, Oberstenfeld, Klienbattan, Winnendon, Eisslingen and Stettin.

April 23rd we were relieved from the 100th Division and again attached to the 103rd Division and continued at a fast pace.

We crossed the Danube River on April 26 and took off for the Austrian border. We were in town after town and did not find resistance anywhere.

German soldiers were giving up by the thousands and thousands. We just told them to keep walking to the rear. They lined both sides of any road you went down, walking single file. None of them had weapons. Some seemed in good condition, but many were in bad shape, wounded, etc.

On May 1, 1945, we got orders to come out of combat and go back to Kaufbeuren to an airfield to guard prisoners of war.

That day we had completed our 240th consecutive day of combat. During this time we fired 73,062 rounds of 105mm shells, an average of 304 rounds a day. There had been days we didn't fire at all, and other days when we fired way more than the average.

I saw Sergeant Eller at the Message Center. He was still writing commendations for medals as we were being pulled off the line to help guard the hundreds of prisoners literally covering the countryside.

We shook hands, and he said, "Well, Walsh, you were right! How much longer do you give it now?" I said, "Three days to a week."

No German soldiers wanted to die anymore and there was no resistance anywhere except a few individual fanatic SS. The SS either fought until they were killed or they killed themselves.

A few days later, all combat ended in Europe

("Germany had lost over four million killed in the service between 1939 and 1945. This is KILLED. It doesn't count the wounded. Almost six hundred thousand civilians were killed, not counting wounded. In this "lost generation" the women outnumbered the men by one fourth. There were probably another million Germans killed or died of exposure or starvation that fled ahead of the Russian Army in 1944 and 1945.

The Russians did not allow a single German to stay in any of the areas that Germany had occupied in previous years and had settled with German families. As the Russian army raged through those areas, any German stragglers they found were shot or hanged in complete vengeance. Women's hands were nailed to wooden carts, the women raped, and then killed. Fourteen million escaped to the American and British zones.

The Russians lost seven million dead in the military and seven million civilians so they had a great hate for the Germans.

The Polish lost twenty per cent of their population."

---- THE SECOND WORLD WAR by John Keegan. See other data in the epilogue of Keegan's book.)

The last few days of the war we were chasing Germans as fast as the crowded roads would allow and they were surrendering by the tens of thousands. The Supreme Allied Command under Eisenhower believed there would be a last ditch stand by the Germans once they got back in the Alps. They also believed through Intelligence that the Germans had vast stores of ammunition, equipment, food, etc, up in the mountains, even poison gas. With them having the high ground they could possibly cause the war to go on for several more months.

This supply theory proved not to be true. There were some supplies and even poison gas back in there but not that much. Besides that, very few got to the mountains, because we pushed them too fast, bombed, strafed and shelled them until many of the outfits were in utter confusion and gave up.

Most of the common soldiers were glad it was over, for them. I don't know of any that caused our group any trouble. Most that we tried to talk to didn't care to talk to us or said German was all they understood, but a few would try to talk in broken English, or in German. They said they were glad it was over and they were alive. They were curious to know what was to be done with them. We tried to tell them they would be prisoners of war under their own officers for a while until we could process and discharge them out of the German army. Then they would get to go home but it would take most of the summer because there were so many. We said we would feed them and have guards around and guards traveling around in Jeeps checking on them to see their officers had them under control.

German officers got to keep their side arms and were actually in control of their men. They were very obedient, no sass or back talk. But they got very quiet when we would come in to their campsites to check with the officers to see if everything was going okay.

One day there was three of us in a Jeep and we drove to a camp and the German Lieutenant came running, saluted, and I said, "We're just corporals, you don't need to salute us," and he said, "I know you are corporals, but you are the conquerors."

As we checked houses, Campbell, the C Battery agent, found an old six-shooter and a box of shells. He was really tickled. Said he always went to cowboy movies and had always wanted a six-shooter but never had enough money to buy one.

So he puts it in his belt. Later as it was getting dark we pulled into a village and took over a house to stay in for the night. Germans were still lining both sides of the road. We had noticed in the fields surrounding the village, hundreds of Jewish concentration camp prisoners roaming around. The guards of a camp had evidently fled and the prisoners had taken off in all directions. Dressed in striped prison clothes, they were mostly looking for food. They were not breaking into houses and taking revenge like you might think or even trying to take German food.

Our little group from the Message Center picked up a couple of them as did many other GIs and one of the two we took in to the house we had taken over spoke English. He was a doctor and the other a shopkeeper. We gave them some food and the shopkeeper was just gulping it down. The doctor stopped him and said, "You can't do that; you have had so little to eat for so long, it will surely make you sick or even kill you."

The doctor translated that the man didn't care if it did kill him, so the doctor took the food away from him, telling him he could have a few bits more in an hour or so until he got used to it. He told him to drink plenty of water. The guy cried like a baby, but he was all right when we pulled out next morning.

We stayed the night in the house with the two concentration camp prisoners. We had just finished eating. There were eight of us and the two Jews, sitting around in one small room. Campbell was sitting at the table and he says, "Now I get a chance to look over the six-shooter I "liberated" today."

So he gets out the six-shooter, looks it all over and says, "I bet I can spin this on my finger just like the cowboys in the movies." He spins it and the damn thing went off.

The bullet went right by his ear and off at an angle through the ceiling, knocking plaster everywhere. Now the German shells have black powder in them and the room filled with black smoke. You couldn't see anything as we had only a couple of candles for light.

Everyone fell flat on the floor and the Jews were yelling, scared to death.

From over in the corner on the floor, Jackson, the B Battery agent says, "Walsh, take that gun away from that dumb son of a bitch. I told him not to fool around with that gun if it was loaded. I said, "Well, I told him to leave one bullet out of the cycling so he wouldn't get shot in the crotch carrying it in his belt." I guess he didn't know there was a war on and he shouldn't be playing cowboys and Indians.

I said, "Campbell, give me that damn gun and I'll unload it." He didn't answer.

In the meantime, Message Center Sergeant Emerton had opened the door and the black smoke was clearing out. Campbell was just sitting there staring, his face black from the oily black smoke.

I took the gun out of his hand and said, “Campbell, are you okay?” He didn’t answer! He couldn’t hear a thing, especially with his right ear so I yelled at him again, “Are you okay?”

We figured it broke his right ear drum or maybe both ear drums. We asked if he wanted to go to the doctor but he said no, just like the time the 50 cal. cap went through his hand.

I unloaded the gun, put the shell in the box with the rest of them and put the box in his jacket pocket. By this time he could hear a little so Jackson and I sat down with him and laid down the rules. He could carry the gun in his belt but there would be no shells in it. If he wanted to look like a big shot, no one would know it wasn’t loaded. If he wanted to shoot it, he had to go out in a field alone and put only 5 shells in it and bring it back empty. He agreed.

Jackson was still mad about it later that night when we stood guard together. He said, “The damn fool could have killed one of us here when the war is almost over.”

I said, “I know, but it’s done now and he agreed not to have it loaded anymore. I don’t think he’ll ever hear out of that right ear again.”

Campbell couldn’t hear very good yet when I saw him about three days after the gun went off. Then I never saw him again.

I never saw Jackson again.

I have seen Emerton several times at Army Reunions in Kearney. He was a rancher near Taylor, Nebraska.

I never saw the Service Battery agent Wagner again. We did exchange Christmas cards for a few years after the war. He was a J.C.Penney store manager in Minnesota the last I heard.

I don't know what these guys did the rest of the time they were in Europe but I suppose they took their turns on road blocks, guard posts, and going on sight-seeing trips like I did.

242nd Field Artillery Battalion

Crossing the Danube

Combat – Spring 1945

CHAPTER TWENTY

As combat ended, anyone could capture Germans by the dozens as they came out of everywhere to give up. We didn't even search them, just said, "Keep on going to the rear. Someone further back will take care of you." We were moving too fast to bother with them.

First it was dozens, then hundreds, and then thousands. You couldn't believe there were that many Germans in front of us. We would just get a hold of one who understood English and tell him to tell the rest to keep moving to the rear.

There got to be so many in different areas that we had to stop and start hauling them back in trucks so they could be fed and kept warm.

At one time before the ceasefire we had over five thousand on an airfield. A fine rain mixed with snow was coming down. We set up guards with machine guns on each corner of the field and held them there. We gave them K rations. That's all we had. We gathered what wood there was around to make bonfires and held the prisoners there both day and night for a couple of days until trucks came to haul them further back.

We put 50 to 60 in each 6x6 truck. They had to stand and could hardly move they were packed in so tight. We finally got rid of the last of them and moved up behind the armor again. The war was about over.

On the last day, a whole German army gave up in front of us. We stopped to take control of them. We let the officers keep their guns and guard their own men, get them camped in different areas to wait to be processed and discharged by us to go home. We had no problems with them. The officers kept things under control and were very cooperative.

There was a German soldier among those on the airfield who was from Iowa. He was an Iowa farmer, had gone back to Germany in the 30's for a visit and got caught there when the military buildup came and was put in the German army. He was a big help to us in holding the men on the field when it was so cold and wet, and not enough food or fuel for fires. He talked English like any Midwest farmer and could speak German to the other prisoners.

In the middle of the night it was raining and snowing. Another guy and I had guard post on one corner of the field. One of the prisoners went off his rocker, started screaming and ran off into the darkness at the edge of the field. The Iowa guy cried out, "Don't shoot, don't shoot!" We did not, would not have anyway. We let him take two others and go get the one that went haywire. When they brought him back, they said the guy was just so cold and hungry and scared, was all that was wrong. They had a few little bonfires going, nothing much to burn.

I walked into this mob with him to one of the fires and made the others move back enough to get this guy by the fire, and had them stand to shelter him from the wind. By now there was a steady half rain, half snow coming down. All the GIs standing guard had used up all the K rations we had this night, so we had no food for him. We watched him for awhile and he went to sleep. Some of the Germans said, no he was dead, but the Iowa guy felt his pulse and said no he was sleeping.

Next morning I had him and the Iowa German put on the first truck that came in to haul them back to a camp.

This Iowa German soldier cussed worse than any American GI I ever heard. He told me he had been afraid he would be killed and never get back to his family in Iowa, but now he felt he might make it through alive, although he was concerned he could not make American officials believe he was really from Iowa. I told him they should be able to check it out easy enough but it would probably take some time.

He cussed Hitler, Goering and other Nazi officials. He said he hoped they all got hung.

During the time I was in the midst of the prisoners, the other guard on my corner was ready with the 50 cal. machine gun in case things went wrong. It was mounted on a circular track on the top of an open cab truck. You just stood in the cab and could fire in any direction.

When this bunch was all discharged down to the officers, the officers had to turn in their weapons. The General at the head of the army turned his Luger pistol over to one of our corporals because he had grown to like him so much. Actually they were supposed to turn weapons over to an American officer of equal rank.

All the prisoners I ever helped transport to the rear did not have to have a gun held on them, or be secured in any way. I suppose we trusted them too much, but we never had any trouble.

Had another prisoner one time Wagner and I were going to take back. We put him in the right front seat of the Jeep and Wagner was going to drive. I was going to sit on the tarp-covered seat in the back. Now it had bedrolls and duffle bags under this tarp and I could not quite crawl up on the big pile of stuff. The prisoner kept saying, “Bitta, bitta” (please), I said, “What the hell do you want?” and he got out of his place and gave me a boost up.

Wagner says, “Well, maybe we should let him drive, too. He’s so damned agreeable.”

The only prisoner I ever captured myself was in a town we had just taken. We didn’t think any Germans were there anymore and we were going place to place looking for loot or booze. Should have been more careful, but I went in a bar type rooming house, not carrying a rifle, nor did I have my pistol at ready.

Some woman says, “Don’t shoot, don’t shoot.”

I didn’t know what was going on but here was a German soldier on a cot. He was wounded but not bad. The rest of his unit had gone off and left him.

I checked him out and he did not have a gun so I went to the door and called our medics. They came and got him and took him to the rear in an ambulance.

I didn’t think much about it at the time but later decided to be more careful at the next town we went into.

We pulled into a grassy place, beside a lake. We set up our tents and kitchen, eat, then put out guards and the rest of us went to sleep.

At sunrise, we lined up the cannon, hub to hub and unloaded all the ammunition and other equipment off the trucks near the cannon. No more danger of air attack.

The kitchen equipment was also taken off its truck and all the big trucks and some of the Jeeps and weapons carriers took off for the air base. Assistant drivers and some other guys went along to do guard duty or to help load prisoners on trucks. We had thought the five thousand on the field we had handled before was a lot of prisoners but it held, we guessed, maybe fifty to seventy-five thousand. (Heard later it was more than that.) Of course some were coming in all the time and trucks were hauling some out all the time. Miles and miles of big trucks, bumper to bumper. Drivers blowing horns and swearing.

Some of us started helping load them. The guys who had been loading them said, “You got to force them to get as many as you can on a truck because there is no food here and some haven't eaten for several days. They need to be shipped back further where kitchens are set up and there is food for them." It got to be a contest to see how many you could get on a 6x6. The record so far was 85 men standing, with the tailgate safety strap across the back. A kitchen truck had a slightly longer box on it and we got 97 on one of them. They couldn't move. There was no danger of falling down, because they were packed in so tight. They cooperated well, and we kept saying, "Essen gayen" over and over to them (going for food).

I never went as an assistant driver on any of these trips so I'm not sure how long they had to ride this way. The drivers said it was usually four or five hours. They were not supposed to stop or let them get off to go to the toilet and quite a few wet their pants or did a pile in them, a few collapsed, but I never heard of any dying. They were turned over to medics and the report was they were weak from lack of food and would be okay in a day or two.

The day after the war ended Staff Sergeant Raymond Pfander (San Francisco, California) and one of the cooks were out souvenir hunting. The Sergeant had come to the 242nd in the original cadre from Camp Roberts that started the battalion back in Oregon. He had moved up through the ranks from Tech. Corporal to become Staff Sergeant of A Battery Supply.

He was not a nasty big egotist like many sergeants got to be. In fact, he didn't fit in with most of the non-coms and most of the time associated with the privates and pfcs. He ran around with the same guys I did in Washington and Oregon. His wife and two little girls came up from their home in San Francisco to see him when we were in Camp White and I met them one weekend. Cute little girls, probably 3 and 5 at the time.

I saw them again when we were in Fort Sill. They had come to Oklahoma to live while we were at Sill.

Strange thing to remember, but in Camp White I remember he had gotten a gum disease and lost all his teeth. The army had made false teeth for him.

I talked to him many times in combat. He didn't seem too worried about the war and was getting letters and pictures from home regularly.

As he and the cook were hunting for stuff along the road, he was killed by a sniper.

I could never get anyone to tell me if the cook was wounded, if the sniper was caught. The whole thing seemed strange to me. No one who was there would talk about it. I have asked at reunions and no one will admit they know anything about it. Next time I see someone from the 242nd I'm going to ask again. Maybe by now they will talk.

I was at the air base a few days and then had a day or so off at the lake where we had parked the equipment. We bathed in the lake, went out in rowboats, or just sat around, or slept. It was real nice there.

Every time any of the trucks came back to camp they brought a load of German small equipment. We had piles and piles of rifles, picks, shovels, dispatch cases and other things like machine pistols, grenades of the potato masher type, ammunition, etc.

A German rifle was a bolt action and the machined steel parts, barrel, bolt, etc., were real good and the rifle was real accurate to shoot. The Germans never developed or maybe never tried to develop a semi-automatic rifle like the American M1. The stock on the German rifle was never finished; it was just rough wood, not even oiled. (They were short on oil.) It looked like someone had whittled it out with a jackknife.)

John Murray (Sydney, Nebraska), Rubin Kesselman (Baltimore, Maryland) and I were out on a rowboat one day. Here came 2 American P47 planes down the lake, just skimming the water. They went by us waving. About a mile further down, one got his propeller in the water. The plane went end over end in the water. We figured the pilot was killed. The other plane went up higher and kept coming back in looking, finally he took off and went out of sight. This happened across the lake and about a mile or so down. We heard two stories the next few days, one the pilot lived, and the other that he was killed. Don't know which was true. One thing we did know, if he lived he probably was in a lot of trouble for losing that plane.

A few days went by. It didn't get around for me to take another turn at the airbase again but our trucks kept going night and day with drivers and assistant drivers and probably, assistant, assistant drivers. Then the trucks were back. The prisoners had all been hauled back and spread out in other areas. Food, medicine and tents were coming in to take care of them locally. There were literally hundreds of thousands around Bavaria, mostly under control of their own German officers.

One morning quite a few of us were told to go with the trucks and start rounding up "DPs", Displaced Persons. We were to haul them to train depots where they would be loaded on cattle cars to be shipped back to their native country. We did not have trouble with the French and the lowland countries like Holland, Belgium, Luxembourg, or those from Austria and Italy. They all wanted to go home. But we had a hell of a time with those that were to be sent back to the countries controlled by the Russians, especially the Poles, Czechs, Latvians, Estonians, etc.

These folks gathered around in bunches in towns and on farms and we had to haul them in at gunpoint, had to search and disarm some of them. When we got them to a train, we couldn't get them on the train. They kept screaming, yelling, crying and swearing at us. We fired in the air, and the infantry guys helping us put bayonets on their M1 rifles.

The infantry prodded the DPs some with the bayonets. When some of the men DPs got real rough, the infantry popped them with a rifle butt, picked them up and threw them in the boxcars. When the boxcars were full, the doors were closed and bolted so they could not get out. There were buckets in the cars for them to use for toilets, and cans of water but no food except what they might be carrying. Probably about 150 or more were put in a car.

I always wondered what the Russians did with them, probably killed or starved a lot of them. There were men, women, children and babies. They had been slave laborers for the Germans.

There were so many men in trucks running around the country picking them up that we got a long train load the very first day and sent the train on its way, probably 20 or 30 boxcars.

The second day word had gotten around and the DPs scattered quite a bit, but we got about 15 or so car-loads. By the third day I don't think we got three car-loads so this action was called off. It would be started again when they bunched up again and then there would be more raids to pick them up.

We went back to the camp at the lake and wondered what would be next. We didn't have long to wait.

Ammersee after Combat ended May 1945

Joseph Freas, Santa Clara, California

Oliver Smelcer, Pattonsburg, Missouri

Anton Striz, Riverside Illinois

Victor Lutkemier, Campbell, Nebraska

CHAPTER TWENTYONE

First Sgt. Reed approached me and said, "Kid Walsh, I got a job for you. The OSS needs three sets of armed guards to back up the German civilian guards at the prison at Landsberg Am Leck (on the Leck River). I have two other corporals and 12 PFCs to go with you. You will be in charge of the group. You will get your orders from the German warden and an OSS guy. Now this OSS guy might be in officer's uniform or a private's uniform or maybe civilian's clothes. Take all your clothes and equipment with you because we don't know how long you will be there and we may be gone from the airbase and moved somewhere else. Someone will bring food and water to you and keep in touch with you every day or so."

I don't know how long the 242nd stayed at the lake or what they did during the time we were guarding the prison. I wondered what they did with those piles of German equipment that had been gathered up. Was it destroyed? Was it hauled to some big collection point? I don't know.

Early the next morning, we took all our personal belongings and loaded onto a big 6x6 truck for Landsberg. It wasn't far. We drove along the Lech River the last few miles. It was about a hundred fifty yards wide and looked to be real deep. The water moved real fast, a lot faster than the Rhine and other rivers we had seen. It was real clean and bluish green. At the edge of the town there was a strange looking hydroelectric plant with a dam across the river, and a bridge built just above it that crossed the river. I had never before, and have never since seen a setup like this and I would like to know how it worked.

About a mile further we came to the prison. The front gate had a castle-like structure on both sides of it, both about two and a half stories high, made of stone and bricks with red tiled spires on top of each, and from these a wall about 20 foot high ran around the whole prison. Driving by it, you would believe it was a big estate with all the trees along the street and large fancy houses with trees and shrubs lining the streets.

The cell blocks inside the walls were separate buildings, one long building about 50 or 60 feet wide and 2 wings running off of it at right angles. The courtyard was planted to small trees, shrubs and flowers, all well trimmed and cultivated. There were lots of sidewalks and a few small statues.

When we drove up to the gate, an American Captain and a corporal and a German civilian came out to meet us. I saluted the Captain and said we were reporting as ordered. He said, "First I'll tell you all about this place and then we will go on a tour and I'll show you where the guards will be and also where they cannot go."

He began, "This is the prison Adolph Hitler was in, in the 1920's and where he wrote "Mein Kampf". Rudolph Hess was in the same room with him. Hitler dictated the book and Hess wrote it down for him. I know you will want to see this room and the view from his window so we will all go there now. You will probably not go in the main part of the prison again. The guards will be at the gate and out in the courtyard and won't be needed unless there is an emergency, which I doubt. There are a few criminals in one area but most of the rest of them are political prisoners. As quick as we can check them we will release them if they were just anti-Nazi."

He paused, "We don't believe there were mass executions here although through the years there were some hangings and some firing squads. All prisoners here seem to be German so they may have executed murderers or ones they thought to be traitors. It doesn't seem the military was too involved here. As you know Dachau Concentration Camp is only a few miles from here and that is where the Jews, Gypsies, enemies of the State and other factions were sent."

(The GI was a very curious sort and when the guards were off duty we walked around for miles in the town and countryside, always in pairs or more, and found a cemetery not too far back of the prison with a lot of unmarked graves ---so all we could do was wonder.)

The Captain said to put a guard at the gate and have three just roam around the courtyard on each shift. The Corporal of the Guard was to stay in the warden's office and do whatever he wanted - - even sleep on the big fancy couch if he wanted to.

Then he said that the big fancy house outside the wall, just to the right of the gate was to be ours. The owners had been told to get out until they were advised to come back. It was a two and a half story house and was really furnished nice. He said he was letting a couple of DPs use a room upstairs until they figured out what to do with them. (I wasn't too happy about this but didn't say anything.)

The area we got to see where Hitler and Hess had been imprisoned was not really cells with bars as we know them, but thick walled rooms, nice size about 20 by 10 on both sides of a long, long hall that was quite wide. The doors were big heavy thick wood and the windows were barred. All you could see from Hitler's cell was the wall around the prison, probably 40 feet away. There were trees on the other side of the wall and that's all that could be seen. Of course he had been there twenty years earlier and the trees would not have been as tall. This room was on the second floor.

A few days later in the downstairs of our nice house we could smell smoke, and something like onions cooking. One of our guys said, "What the hell's going on?"

We went upstairs and the DPs had an open fire in a metal bucket making soup out of some wild onions they had found somewhere. I said, "You take that stinking mess outside to cook it," and we finally got them to go. I don't know if they were Poles, Czechs, or Russians but some kind of language like that. They had scorched a big spot right in the middle of the polished oak floor.

I said, "This is enough of this shit. Some of us will be sleeping in here and they will burn the place down and us with it."

About that time the Captain drove up to the gate of the prison, only he was dressed in civilian clothes. I went out and got him and showed him what had happened. It still smelled, burnt wood, varnish and onions. He said, "Well there are thousands roaming around here. I guess there will be two more."

So we went out in the yard to the DPs and he rattled off about two minutes of some language and the DPs took their fire bucket and soup and headed down the street. We never saw them again.

We never learned where they interrogated the Landsberg prisoners. It wasn't in the warden's office. Maybe there was another office back of the warden's quarters on the other side of the gate or maybe they questioned them in their cells. But every few days they would bring one to the gate and let him go free.

One day the OSS said, "Don't put any guard on the west wall tonight. We've got something going on and we don't want any mix-up."

The next morning one of them said, "I'll show you where someone went over the wall last night." Sure enough, some of the stucco on the wall was broken off and there were shoe marks on the upper 8 or 10 feet of the wall.

I said, "Well, you didn't want a guard in this area last night."

He said, "That's right, but you can put one back here tonight. I just wanted to show you."

No other explanation. We figured it was some agent or double agent they didn't want any record of, or let him out to follow him or for him to locate someone they wanted who was bigger than him. We never found out any more.

The old German warden didn't talk hardly any English or didn't want to, as was the case with many of them. So we never found out anything from him. He and his wife and a daughter, probably in her early 20's lived in the quarters on the left side of the prison gate. There was one window that looked out on the street on the second story and the family would sit around this window in the evening. Much of the time the girl sat squarely in front of the window with her elbows on the sill looking out about every night for a couple of hours at a time. It must have been their living room with the bedrooms, dining room and kitchen behind. The daughter always spoke when we passed her – in German, "Gooten Morgan" in the morning and "Gooten Abent" in the evening. She was rather pretty with long brown hair and a little on the plump side. I think her name was Gretchen.

Once in a while the parents and daughter would come out to the gate and go for a walk in the evening. They always told us, "Spatzerren geyen" (Going for a walk) so we opened the gate and let them out, and let them in when they returned. This was about all the conversation we would have with them.

I would try to talk to the warden in his office during the day, and he would say "Nix furstay" or "Furstansy nix", but I was sure by watching his eyes and his fidgeting around that he did understand but he never admitted it. I finally gave up and just spoke to him every day and let it go at that.

One evening my guards had already left the house to go over to the prison. I was getting a couple of books and my letter writing stuff together to have something to do in the warden's office for a few hours and was starting over to the prison. As I came up the walk, I noticed the girl and her folks sitting by their window. I kind of half waved and said 'Good evening" in German and they answered in German. I walked on and as the guard was opening the gate to let me in, the girl called down, "Bitta". I was turning around when she said in pretty good English, "Corporal Walsh, could we please speak to you?"

I must have really looked surprised. My suspicions were correct, the dad, mother, and daughter all spoke English. I walked back under the window and looked up and said, "Sure". I never in a million years would have guessed what was coming next.

She said, "Please, could you tell us if you know anything about Oklahoma in the United States?"

I said, "Yes, I spent about a year and a half there."

She says, "Do you know a place called Fort Sill?"

I said, "Sure, that's where I was stationed. How did you know that?"

She says, "Oh, we didn't know YOU were there, we just wanted to know what kind of a place it is and what it looks like around there."

I was beginning to get a little griped. I said, "It's a beautiful military base and the country around it is pretty but I'm not saying anymore until you tell me what this is all about. You pretended you didn't understand English and now I find that you do." I could tell by the parents' faces that they understood everything we had said. "Now what's this really about?"

She began to cry, "My husband is a prisoner of war, and the Americans have him at Fort Sill, Oklahoma." I could have fell over!

She said, "We fear how he is being treated, how he has to live and if he gets enough to eat. I have letters from him and he says he is fine, has a warm place to sleep, a good bed, good food, (she held up some letters) but we think he may just be writing this to keep me from worrying."

I thought, "My God, what have I got into now?" I said, "I've got to go into the prison and make rounds with the guards. Come down to the office in about 15 minutes and I will talk to you about Fort Sill."

As I posted my guards, another Corporal picked up his guards as we walked around the courtyard. We were all talking about this and everyone was saying how they always thought about 50% of the population understood and spoke English. We all thought it would be all right to tell the family anything they wanted to know about Fort Sill. After all the war was over. Practically all of us had guarded German prisoners at Fort Sill and also Camp Howze, probably even her husband. But we couldn't figure out how they could have come to believe we had been at Fort Sill. I decided to ask this first and then tell them about Sill.

They were already standing in the office when I got there. The girl apologized for the intrusion and for causing me any trouble. I said, "It's no trouble. First I want to know how you knew my name and that I may have been at Fort Sill."

She said they had heard the other soldiers call me "Walsh". She said they had a pamphlet on American uniform identification and knew the red piping on our caps and the crossed cannon buttons on our lapels were field artillery and in some book (probably encyclopedia, she didn't have an English word for it) they had found Fort Sill and knew it was a big Field Artillery base that had schools for training officers and men in all phases of artillery and the Fort had been there since the Indian wars. (All Europeans were fascinated with anything they could hear about Indians.)

"What would you like to know first?" I asked. She said she wanted to know about food, shelter, and medicine.

I told her the prisoners got the same food the soldiers eat, it is good, there is plenty of it and they get three meals a day. There is pork, beef, poultry, fish, vegetables, bread, cakes, desserts and even ice cream. I told her the prisoners were paid a small wage every month and they could buy cigarettes, cigars, toilet goods, candy, etc.

I told her about the temporary wooden barracks with cots, blankets, pillows, also lockers and showers just like the Americans had. In fact, some of the Americans lived in tents and that I had myself. The barracks were heated in winter; there was no shortage of fuel at Fort Sill. (I didn't tell them about Yakima!)

I said that the prisoners were treated by the same doctors that took care of the American soldiers and that the prisoners had medical checkups once a month just like the GIs.

She seemed relieved to hear all this and asked, "Do the men have to stay in that barracks all the time?"

I said, "No, these barracks are spread out over a large area. There is a high fence around it with two American guards on each corner, two at the gate, and about twenty or so more in a guardhouse outside the gate. The big yard between the barracks is green grass and the prisoners play lots of soccer and other games out there. They dress in just shorts and shoes when it is warm and they look like they have been on a beach all summer. They also have games to play in the barracks."

"Every day small groups are taken by guards to different places on the base, in trucks to work. They clean and repair streets, haul garbage to a dump, mow grass, trim shrubs, clean buildings like the theater, work in kitchens cleaning, washing dishes, setting tables, some serve food. They do many other jobs and when evening comes they are returned to the prison compound, they shower and eat and then play soccer until dark, or just sit around, rest, write letters." I told her they seemed satisfied with their situation. I was on guard there two times on a 24 hour shift but I never talked to any of the men. I did see all of them as they went to work and came back so I probably saw her husband as she had said he went there early in the war. I asked her if he had been in North Africa and she said yes.

She asked what the area around the fort was like and I told her it was hot in summer, cool in winter but not real cold. The farmers raised cotton, corn, and wheat and the ranchers raised cattle on the grasslands. I mentioned the few mountains on the base, nothing compared to the Alps, a few small streams, lots of brush and trees, and lots of very big pecan trees. I told her the town of Lawton was probably about the size of Landsberg. Oklahoma is about one half the size of Germany. If you added Kansas, then it would be about the size of Germany.

She knew the United States was large but hadn't thought of it this way. As we were talking, the family talked among themselves in German. The parents never did say anything in English or admit they understood it. They continued this way all the while we stayed at the prison. (To this day I think they understood every word.)

I asked if there was anything else I could tell them. She said no but maybe later, and thanked me in English. The parents thanked me in German, "Danka, and Dankashein." They all took turns saying good night and left. Over two hours had passed.

As the days went by, the daughter did talk to me and a couple of the other guards, but not too much and always from the window in the evening. She occasionally had a question or two. She asked once how soon I thought her husband would be sent back home. I told her probably in a few months because the big ships taking Americans home would be bringing Germans and Italians back on the return trip and that most of them should be home by Christmas. I often wondered when he did get back.

One of the wings that extended to the right of the main building housed about 25 or 30 women prisoners. Although we never went into these wings, you could see that the 2nd story was just one giant sized room and the women all lived together in this room. There were continuous sections of barred windows the full length of the wing on both sides. About half of the women appeared to be between 20 to maybe 35, and the rest on up to 60 or 65.

The first day we arrived the younger girls started calling down to us, making remarks in German that of course we did not understand. We thought they were heckling and making fun of us, so we just waved at them and then ignored them. They evidently worked during the day for this always took place in the evenings.

After a few days one of the GIs started trying to talk to them and found one who could speak fairly good

English. They knew we were Americans and claimed they were never told the war was over. This guy was from Oklahoma and was pretty mouthy, and right away he accused them of heckling us. They said it was not that way; they just wanted to know about us, where we were from, how old we were, and so forth.

The English speaking girl had a heck of a time with the rest of them, jabbering at her to ask this and that. Then they would start laughing among themselves and he would say they were making fun of the Americans and that we would not talk to them anymore. She would repeat this to the others in German and there would be a big moan go through the bunch of them.

One evening when I took my guards around, the girls were already at the windows for us and started calling down questions right away. I said, "Wait a minute! How about we ask the questions, how old are you? What are you in here for?"

It seemed they were all political prisoners, had said something against the war or Hitler or the government in public and had been reported. They had been arrested, found guilty, and sentenced to prison, most of them for several years. Some had been there for a few months, and some for two or three years, some of the older ones longer than that.

They were from several different areas of Germany and the English speaking one said they had different accents and she had difficulty understanding some of them. There was a very pretty platinum blonde girl about 32 years old who was from Rhineland area that had a different accent and spoke very fast. Most of them could not understand her and could not get her to speak slower. She had spoken out against the war after her husband had been killed in action on the Russian front.

One evening as I checked my shift of guards I found this kid from Oklahoma had climbed up a trellis and was sitting on a ledge of the second story window with all the younger girls crowded around the inside of the bars talking to him.

I called up, "What the hell is going on?" He yelled down, "Come on up, Walsh!" I said, "Are you crazy? I got through this war and I'm not going to crawl up a trellis to talk to girls and fall off of a ledge and break my neck! Now, you get down here or you are in trouble. You even left your carbine leaning against the wall down here."

He came down. The girls booed me. He said he didn't see what was wrong with going up there. I said, "I don't care what you do when you are off guard duty, just what you do when you are on guard." So after that he would climb up there when other guys were standing guard.

One time I went around by the windows and he was up on the ledge holding hands and necking with the 22 year old through the bars. I don't know if they tried anything else through the bars after dark or not, but I wouldn't doubt it for a minute. She had quite a crush on him.

I always stopped to talk to the girls when I took my guards around to their posts. Some of the others not on guard would go and talk to them too. But nobody else ever went up on the window ledge. The platinum blonde was always having questions asked about me, all kinds of questions, probably because I was older than the other guys. One evening they told me she was being released the next day and she would like to meet me the next evening at the bridge over the Lech River (which was about a half a mile from the prison) about seven o'clock. I said I was on guard at seven but I could try to get someone to trade guard with me. I knew the meeting time could not be later because of the curfew. All civilians had to be inside before dark.

The next morning she and three others were released. The driver for the OSS officer loaded them in his Jeep and drove away. I was walking over to the prison about this time. As they drove out, the girl waved. That evening another guy and I started for the bridge where I was to meet this girl. We always walked in pairs because we were still concerned about the possibility of snipers.

As we neared the bridge we decided Howard would wait about 50 yards away and I would go meet her. We were suspicious, but curious, too. Maybe she did want to just meet me or maybe it was some kind of trap. We couldn't see any reason for that. We were a little early but I went about a third of the way across the bridge and walked up and down and back and forth. Didn't want to be a motionless target.

It got to be about 7:20 and the only activity was an occasional civilian on a bicycle. I waved to Howard to come to the bridge and we decided to cross to the other side, maybe we had misunderstood the location. No one anywhere. It was now past 7:30 and we decided to head back. Here came a boy about 10 or 12 years old calling, "Allo, allo, bitta, bitta!" We stopped and he had a note for me, - - - but it was in German.

When we got back to the prison, the warden's daughter was sitting in the window as usual and we asked her to come down and read the note for us. It read, "Could not get there tonight. Come to the place I am staying tomorrow night." It had the address and was signed "Gretchen". We asked where the address was and it was clear across town. Another Gretchen.

The next evening four of us headed out. I had given up the idea of a date with this Gretchen but we were curious. We were already wondering if the boy could be her son. We crossed the river bridge by the power plant, went about a mile through a factory area along the river, then a residential area and then came to the center of town. There were quite a few civilians on the streets and we got a lot of nasty looks from some. There were some German soldiers in uniforms yet. They didn't look too friendly, either.

We were about halfway through the center of town when we saw a beer hall open so we went in and had a beer. We had occupation money to pay for it. We had all been issued occupation money before we crossed the border into Germany.

It got very quiet when we entered and stayed that way while we drank our beer. There had been voices and laughter coming from the hall before we went in. There was one corporal and two privates with me and the corporal said, "I believe we had better get out of here. Even though I would like some of that good beer."

I said, "Oh, hell, let's have one more." He said, "Okay, but I don't know if I'm going to enjoy it." We ordered one more and we drank fast and left. We gave the barmaid a nice tip and she did thank us.

When we got outside, one of the privates said, "How far is it to that address?" I could see he was worried. I said, "I don't know but we are not going to try to find it. We are heading back."

As we walked back, Howard said he was more scared than when he was in combat, and if something had happened in that beer hall we would have been in TROUBLE. I said I didn't really think any of those in the hall would cause trouble unless some nut went off his rocker.

Next morning I was talking to one of the OSS people and he said they had all the paperwork done on the people in the prison and had released most of them. He said the Germans would do their own guarding now, that a truck was coming from the 242nd to pick us up, that we should get packed up.

Some of us went around in the courtyard to say goodbye to the girls. There were only 3 younger ones still there and not many of the old ones. The English speaking girl was still there and she cried. The Oklahoma kid said he would keep in touch with her. (And he did. She got out. He re-enlisted for another tour of three years in Germany and had 30 days leave back in the States. They were married when he came back to Germany.)

The 242nd truck came for us. The warden and his wife and daughter came out and told us goodbye and wished us good luck. Gretchen gave me a hug. I never knew when her husband got home.

I was never back in Landsberg again. I always wondered about the blond girl and the little boy. I never understood the OSS operation going on there, letting prisoners escape, the different uniforms and civilian's clothes of the OSS guys. They were always very happy acting and joking among themselves but yet very strange acting. I think they were half nuts from living on the edge and yet loving every minute of it.

These stories about the OSS and the two Gretchens and everything actually happened. None of it is fiction.

Edward Walsh, Gresham, Nebraska

Calvin Howard, The Dalles, Oregon

Landsberg Prison Gate

Landsberg Am Lech, Summer 1945

Landsberg Prison

Landsberg Am Lech – Summer 1945

The cell where Adolph Hitler had been held prior to WWII is marked above the window.

CHAPTER TWENTYTWO

We loaded onto the 242nd truck. The driver was alone and he motioned for me to get in the cab with him. A Battery had moved from the lake into a nearby town but would be there only a few more days. There was a note for me from Sgt. Reed.

"Kid, we need to set up a road block for a few days before you come back to the Battery. Just to check everyone's traveling papers who come along the road. Check their identification papers and picture. If the picture and description don't fit, detain them. We'll have a truck coming by several times a day to pick them up. You have the right to search luggage and their person if you think it necessary. Also you may search any buildings and houses in the area. Confiscate any firearms. They cannot have a knife over four inches long. If you find one, break it and give back the pieces." Signed First Sgt Reed, Battery A, 242nd FA Battalion

We soon arrived at the spot. It was just a small building by a tree along side of the road. Not even a crossroads in sight. The nearest village looked to be 3 or 4 miles away. We were probably about 20 or 25 miles southwest of Landsberg.

We had gotten there in a hurry. These truck drivers were as crazy as the OSS. They knew only one speed – wide open—on these narrow roads. Normally the governor on these trucks would only let them go about 45 or 50 but when we got to France and into combat, the drivers and mechanics had either removed the governor or wired it wide open. They were burning 100 octane aviation gas in the trucks and they would do between 60 or 70 miles an hour. So the European kilometers whizzed by pretty fast. Anyway, that's the way they all drove, especially when there weren't any officers around. I should say, SOME officers. About half of them didn't care.

The driver dropped us off by the tree and roared off down the road. He had brought food, water, candy and cigarettes for us, a couple of single burner gas stoves, and a "jerry" can full of 100 octane gas. The little stoves were supposed to burn white gas, it said so right on the stove. But where the hell could you get white gas? Anyway, they worked fine with 100 octane. It really burned hot. You could boil up a tin can of coffee (about a quart and a half) in just a few minutes. When the stoves leaded up, we used a wire brush on the burner, held it upside down and beat on it to get the carbon out and it would work again for a while. Sometimes we just threw the stove away. I bet there were thousands of them laying around over Europe by the time the war ended.

Also it's a wonder they didn't blow up, like a Molotov cocktail. The Jerry can was a German gas can painted desert tan much like our own five gallon gas cans that were olive green, except it was measured in liters. The German soldiers were called "Jerry" mostly by the British, but also by some Americans. (Most Americans called the Germans "Krauts"; a few used the French term "Le Bosche" pronounced Le Bush.)

We put our gear in the building and cooked up some food and set up the roadblock. There were quite a few people on the road, mostly on bicycles, but some walking. A few with horse drawn carts or wagons.

We were there three and a half days. I don't recall any German cars or trucks on the road. We checked the papers and searched the luggage of several hundred during that time. Never had to detain anyone. We did confiscate several pistols and military type knives. We broke quite a few knives that were over 4 inches long.

A girl came through the checkpoint carrying a large fancy military award type knife. She claimed it was the only keepsake she had of her husband who had been killed by Americans. It was such a tough knife we had difficulty breaking it. She got so mad we had to get a couple guards out to hold her. She was swinging her fists, kicking and screaming, mostly at me and another GI for breaking the knife. Then she swore she would kill me and the other guard some time when we wouldn't expect her to be around.

We watched out for her pretty close the days we were there. She was really violent. She spoke pretty good English, but called us every name she could think of in German. I took a picture of her and she really got mad at that and tried to get the camera. We told her to get on her bike and get going. The guards had to restrain her again. Finally I told her to get on down the road or we would arrest her and send for the military police. She left on the bike, looking back over her shoulder yelling insults and "Ill kill you, I'll kill you!"

She was dressed in a military outfit. I don't know what they were called in the German army. She did have military discharge papers that were signed by an American officer.

(In 1947 when Jeanne and I were staying for a short time with Jeanne's parents and I was driving from Shelby to Seward to work, I thought I saw this girl walking up out of the ditch in a heavy fog one morning just east of Rising City on the north side in a low area. There was no house real close to this spot. The clothes and cap seemed to be the same. To this day, I don't like to go down that road when it is foggy or rainy. --- Surely, after all these years she won't catch up with me and carry out her threat!)

The last day we were at this road block, a DP came in across the fields all excited and said there was a bunch of German women soldiers staying in the loft of a barn about three quarters of a mile across the field. Well, we knew there wasn't really any women combat soldiers in the German Army, so they must be clerical type like American WACs. Three of us went over to take a look.

There were over 20 girls in that loft. The place was all cleaned up. Their bedrolls were lined up on the floor in two neat rows the full length of the loft with their personal belongings in a bag at the head of each bed roll. The girls were just sitting around talking.

An older woman came to meet us but she didn't talk English or didn't want to. She had some papers she gave us that said the Americans knew they were there and they were to stay there until processed out of the service. We asked if anyone spoke English and one girl came forward who could manage some questions and answers. We told her to tell the older one who seemed to be in charge that we were going to check the personal belongings of a few of them and also the hay in the loft for weapons. We checked a few bags and also the hay at both ends of the loft and found nothing. We thanked them and left.

We hadn't even gotten a dirty look from any of them, but as time went on, we found that the German military was always willing to cooperate. They really knew how to play the part of a good loser. Practically all the dirty looks and remarks came from civilians. After a couple of months most of the civilians were real friendly and would talk with us and laugh at the crazy Americans and the things they said and did. The word was something like "trielict" meaning crazy.

They even got so they would use the double meaning of a German word to answer. Like if we would say "Vaas ist loes?" meaning what's going on, or what's new, or what's up, they would say, "The haunt (dog) is loes in der lane," from an old nursery rhyme. They thought this was a big joke.

(Incidentally many of the old nursery rhymes originated in the area and to the north, like the city of Hamelin and the Pied Piper.)

Sometimes they would ask US what was going on! We would say, "Feel (lots of) beer drinking with the Burgermeister (mayor)." At first this really impressed them because the Burgermeister was really a big shot in the villages. They soon found out we were joking and called us crazy.

However we did have contact with many different Burgermeisters because we told them what we wanted done in a village, what houses we needed to live in and so forth. It was up to him to get the people out of the houses and see there was no trouble in the village. He really had a heck of a job to do. We were supposed to put the burden on the Burgermeister. In combat we had just walked in with guns drawn and told the people to get out. We didn't care where they had to go, or we moved them off into one room and took the rest of the house.

Once several of us were sitting under a tree and a girl walked by. One of the guys whistled at her and she turned around and said, "Americanish schist kopf!"

He says, "What did she say?" I knew this word from back in Nebraska and I told him, "She just called you an American shit head!" - - - - So for quite a while after that the guys all called him shithead in German.

The truck finally came to relieve us of the roadblock. The driver said the battalion had taken over five villages about 15 miles south of Munich and we were to be there for a while. A Battery was in the village of Assling not far from a bigger town Rosenhiem. It would be great to get back with the rest of the guys and maybe loaf around for a while.

Assling was a farm commune type village. The people farmed the surrounding land. The village was set up in an oval with the center kind of a park area, grass and a few trees. Reminded me of the infield in some race tracks. Our trucks, cannon and trailers, weapons carriers and Jeeps were lined up in the center of this park in perfect alignment, all cleaned up, even the paint had been touched up.

Each section of men had a house of their own. On the north side of the oval were the four gun sections, each in their own house, a house with the First Sergeant, battery clerk, mailman, supply sergeant, and supply corporal. On the south side were the wire and survey sections. The mechanics, gun mechanics section and the truck drivers also had a house. They had always stayed together in a tent in combat, not with their real section. Also the machine gun and ammunition section had a house. The cooks had taken over a sort of town hall building that had tables and chairs, and had set up a kitchen there and lived in some small rooms also in that building.

It looked like a nice place to spend some time. I moved in with the wire and survey section.

Next day we washed and cleaned up all our clothes. Some that were damaged or mildewed were turned in and replaced with new. The wool uniforms were washed in gas, then hot water and GI soap and hung out to dry. They would be pressed the next day. Boots were washed, polished and shined. The belt harness and musette bags were washed and dried. The brass uniform buttons, belt buckles were shined and of course the carbine and clips were cleaned and shined. This was really the first chance we had to get everything cleaned since we went into combat in August 1944. It was now June 1945.

There were still a couple of guard groups out somewhere from A Battery. One of them that I heard a lot of stories about came back two or three days after we did. They had been guarding a place they called the "Hatchery". It was a large compound where nearly one hundred women were housed who were supposed to be pure German or "Aryan", to have babies from so called pure Aryan soldiers, mostly SS. These guards said they had never seen as many blond blue-eyed women, children and babies if they counted all they had seen in their lifetime. They said it was unbelievable. They had living quarters, hospital, kitchens, and so forth for mass production of babies who were taken by the State at an early age, especially the boys, to be trained as soldiers. (The government had also paid regular families for having babies, more money for a boy.)

The big problem had been that the Americans in charge didn't know what to do with this place and other places like it all over the country.

First was the problem of feeding them. The German government didn't exist any more. There was no German transportation to haul foods. There was no money to pay for food or medical supplies. Most of these women and children had no place to go and many had no relatives. Of course they didn't know where the different fathers were, who may or may not have survived the war, were killed or taken prisoner somewhere. There were tons of records on the women, fathers, children, where the older children had been sent and so forth but it would take months, even years to sort all this out.

The girls who had more than one child, probably had a different soldier for each and some of the men may have been married. It was going to be impossible to locate the fathers. The times we thought about this we thought the thing to do was to put the children in a giant orphanage, go through the records, and attempt to match up the mothers and their children. Girls who had one child and could not have another had been sent on their way and now would have to be located.

I wonder how this mess ever turned out.

One of the men from A Battery while on guard there fell in love with one of the girls who had a baby about 3 months old. He reenlisted to stay in Germany, took his 30 days leave back to the States, returned to Germany and married the German girl. He said he didn't know if she had other children. We told him he should find out before he married her, she might have 2 or 3 somewhere in Germany, but he said he didn't care.

Two or three of the guards got involved with these girls, but most of them passed up the temptation. One of the guys from Fairbury, NE told me he was never so glad in his life than when he was able to get out of there. I asked him why and he just said, "Those damn women."

I asked one of the others who had guarded there about it. He said that the last week on duty there, one of the girls had tried to seduce the Fairbury kid every time he stood guard. She knew he couldn't leave the area for 4 hours. She would unbutton his clothes, toss his cap down the hall, unbuckle his gun belt and let the belt, pistol, ammo, and canteen fall to the floor. He would get red as a beet with all the guys watching, and sweat like it was a hundred degrees. But he never touched her and just kept buttoning up his clothes. She would come out in the hall wearing a thin robe with nothing under it. He did not know that the other guards had set him up by telling the girl that he really liked her and wanted to go to bed with her. One of the guys who pulled this told me if he ever got to Fairbury he was going to tell this guy's wife that her husband was the most faithful husband in the United States Military Service. Yes, the guy was married.

One night when I was Corporal of the Guard, Robert Spotted Wolf (Van Hook, North Dakota) got drunk and was shooting up the place. The walking guard and I got close enough to him and took away his gun. We took him to the house where he slept and told him to go to bed. He got out a knife and went back outside in the dark and told the men in his house he was going to kill us. We got the Officer of the Day and called out all the guards. Sixteen of them rounded him up, put him in a room and stood guard on him all night. Next day he was all right. Claimed he didn't remember any of it. He could have been court-martialed if we had pushed it.

The second day we were in Assling, another GI and I were sitting on a bench in the park near the house we were staying in and here comes a pretty little girl about 4 or 5 years old, blue eyes, blond hair clear down below her shoulders. She walked right up to me and said, "Vas du namen?" I said, "Edward." She said a couple more sentences that sounded like questions, but we didn't understand. She said her name was Gretchen.

"Nix furstay deutsch," I told her. She talked on and we thought maybe she was asking for candy or gum. I told her as best I could in German to wait there, I would be back in a minute. I went upstairs in our house and came back with candy and gum and gave it to her. She smiled and thanked me. She never opened the candy or gum. We watched her as she went down the street, turned the corner and disappeared.

We laughed and sat down on the bench again. We didn't figure out for some time that we had just met the best little con artist in Europe.

The next afternoon a few of us were loafing around on the benches and here she goes down the street, over where we were and said "Allo." Some of us tried to talk to her but no one spoke German. There were only about 3 guys in the whole battalion who did and they were all in headquarters battery in another village. We figured she was after candy and gum again so between all the guys around the bench she got a pretty good haul, candy, gum, K rations, crackers, packets of coffee and boullion powder. GIs always had their pockets full of everything you can imagine. (They had gotten into this habit in combat of carrying everything they might need if they got isolated or pinned down for any length of time.)

She was really happy, smiled and thanked every one and went down the road to the east. She had come from the west. She didn't turn at the corner like the day before but went on past to the next corner and turned left. There wasn't much comment among us except what a pretty little thing she was. There were mostly older people in this village and only 3 or 4 children. Of course quite a few people had to move out of their houses to make room for us.

This was only the beginning. Next day Jim Tynan (Lansing, Michigan) and I were sitting on the steps of our house and here she comes from the west again.

"Gooten Morgan, Edaward," and she sat down on the steps as close as she could get.

Tynan says, "I believe you have a German girlfriend!"

I said, "I believe so."

Tynan says, "I've got her figured out. She's a con artist, and she's here to get food. I bet she has been begging for food and none of us understood what she said. She probably picked on you first because you remind her of someone."

I said, "When the hell did you get to be a psychologist?"

He says, "You know what I mean. We got some C rations in the house. I'll go get some. Maybe her family is hungry." He came back with cans of beans, hash, eggs, stew and a package or two of melba toast (Everyone was having melba toast sent from the States.)

He gave the food to her, she thanked him and went down the street east and turned at the same corner she did the first day. Tynan says, "I bet she shows up again tomorrow."

There was probably five or six of us on the benches the next day and here she comes from the east. She spoke to me calling me, "Edaward." There was no room on the bench so she just crawled up on my lap and sat there. I was some embarrassed. I had never held a little girl before. I had held Bill Walsh, Buzz Pratt and Donna as babies (nephews and nieces). I really didn't know what to do for her and could not make her understand English.

Some of the guys laughed because they could see I didn't know what to do and right away one of the older guys said, "Dammit, Walsh, put your arm around her back before she falls off your lap."

She got 3 complete K ration meals that day, a breakfast, a dinner and a supper plus the usual candy and gum. These K rations had canned eggs, stew, beans, crackers, cigarettes, matches, coffee, bouillon, toilet paper and candy. She went east but some further down the street and turned left.

I got some razing about being embarrassed when she crawled into my lap.

That night, in the middle of the night, there was the loudest, strangest noise anyone ever heard. It woke everyone up. All were grabbing clothes, shoes and guns and running outside thinking we had been bombed and were under attack.

Just up a slight hill to the west about a quarter of a mile away was a railroad depot. Two trains had hit head on right by the station. We were trying to get dressed and drivers were getting keys and getting vehicles started. You could see flames and steam through the darkness down by the depot. Terrible screams were coming from there.

We jumped into Jeeps and into the back of 6x6s as others were jumping on trucks and weapons carriers.

Train Wreck – Assling Germany, Summer 1945

Steam was escaping from the engines but no flames as we arrived at the wreck. The engines were a mess but the big trouble was that one train had four cattle cars that were jammed with German soldiers. Quite a few were running around trying to help others. So much yelling and screaming and crying. There were many dead, many with crushed arms and legs, and some with limbs completely cut off.

We were all trained in first aid but we had no medics with A Battery. It was about five miles to a small hospital and about 25 to 30 miles to a big one in Munich. There were no telephones. The guys tried radio but they were all shut down in this area. (Shutting radios down was a mistake and was changed after this accident. Every company or battery removed a radio from a vehicle and installed it in the guard post so there was a network of communication 24 hours a day.)

Someone decided all that could be done was to just lay the injured ones in the bottom of trucks and get them to hospitals as fast as we could

German and American soldiers were working side-by-side loading them, putting on tourniquets with the bandages from our belt kits. Our bandages had sulfa in them. There was so much screaming and blood. Bodies were pinned in the wreckage, some sticking out of the sides and ends of the wreckage. These had to be removed with crowbars, and welders the next afternoon.

We were at the depot most of the next day, taking turns going back to the village to eat. When I went back the second time, I took my camera and got pictures of the wreck, even the bodies still pinned in there. These pictures are in my album.

The Germans that were not injured were taken down to our kitchen and we fed them. They were then hauled off somewhere. I believe everyone was shook up as much or more than if it had been combat. Some got sick and couldn't help. I can't remember how many were killed but it seems like it was 87. And of course, some of those injured surely died.

A truck driver said that as they were headed for Munich going about 60 miles an hour, one of the injured suddenly pulled himself over the tailgate and tumbled out. The truck close behind ran over him. He had been one of those with his leg crushed clear to the hip.

Late in the afternoon we had done all that we could and went back to Assling. Practically everyone collapsed on their beds. Most fell asleep but some could not and prowled around most of the next night until they finally got so tired they went to sleep, some sitting somewhere on a chair.

Next day the little girl showed up again. Same pattern, came from the west, and left going east. This time her hair had not been combed but she carried a comb and handed it to me. She sat on the bench and I combed her hair. Several men were around so she collected quite a bit of food.

About this time the few kids in Assling started showing up at the kitchen at mealtime with buckets to collect leftovers. They stood by the garbage pails and held their buckets. All the guys in the battery dumped their leftovers in the buckets. The little girl was there with a bucket. Almost immediately everyone started putting more food on their mess kit than they could eat so they would have plenty to put in the buckets. The little girl was getting the lion's share of the good stuff.

So this started a chain of trouble with the cooks. They were running out of food before all the men got through the line. Besides that, the cooks were bitching they were the only ones who had to work, that the rest of us were just loafing around. They forgot that we were going out on guard posts, hikes, and keeping the equipment up like it was done in the States.

So the Captain had a big meeting out in the park and we got orders not to take more food than we could eat. The cooks were told if they could find four DPs to work for them for meals, it would be okay, but if the DPs wanted money it would have to be out of their own pockets. Well, everyone had a pile of the old German marks they had stolen in combat but it was not worth anything. Only the occupation money could buy anything, but the DPs didn't know that. The Captain laughed when he said this, but sure enough, before the day was over the cooks had hired four DPs, two girls from Austria, and two men from Czechoslovakia. Actually they probably knew the money was no good but were glad to get three meals a day. They did about the same work as a GI on KP in the States would, wash pots and pans, clean up the tables and floor, dish out food and so forth.

It did cut down on the food the men were taking out but they still slipped bread, biscuits, dried fruit, an occasional apple or orange under their shirts to give to the kids. We were beginning to get fruit from Spain, Portugal, and also from Florida.

One of the Austrian girls could talk fairly good English and she had her mouth going all the time. One of the things we all got a kick out of was whenever she was dishing out beans in the mess line, she would tell every guy that they should drink lots of water after beans, at least six glasses. They would ask why, and she would say for their stomach. They would say later when they left the mess hall that she didn't want them all to be farting at the same time and stink up the town.

Tynan got the idea to have her talk to the little girl – but also he wanted a date with the other DP girl. She was blond, and I tell you she was UGLY. The English speaking one was a nice looking brunette, fairly tall and rather plump. Well, she talked to the little girl but didn't really find out anything. She claimed the Austrian German and the little girl's German was different and they didn't understand each other.

But he wouldn't give up on this. He told the Austrian, "Okay, we're going to the Burgermeister and see what he knows about this little girl."

She immediately said, "Oh, no! A DP cannot talk to the Burgermeister, because he is the boss of the town."

Tynan says, "Boss, hell! We, the Americans, are boss here! Let's go."

She went with him and translated. It was strange she and the Burgermeister understood each other but her and the little girl did not.

He found out the little girl and her mother lived with her grandparents, and that the mother was sickly and seldom seen outside the house. The mother seemed much better though than when she and her daughter came to the village about 2 months before. The Burgermeister didn't know about her husband for sure but thought that he was dead, or a prisoner of war or maybe missing in action.

Tynan and the Austrian told us what they had found out and he said he had gotten a date to go for a walk with the blond and I was to go with the one that spoke English. She was still standing there and said, "You'll have to bring some K rations, candy, and cigarettes and we'll have a picnic."

She turned to leave and said, "Do you have any nylon stockings?"

He said, "Hell, no."

She said, "But I heard all Americans carried nylon stockings with them."

Tynan said, “Well you heard wrong But we could probably get a Mercedes and go for a ride tonight.” I think she believed him.

I saw First Sergeant Reed coming down the street and thought, “Oh, what now.” He gave me a paper and said, “Kid, get these six guys and saddle up. You need to set up a guard post at this cross roads right away. There are only two houses there. Take over one of them to stay in. Keep two men at the crossroads at all times. There’s wood there so keep a bonfire going at night so you can see. We’ll get food to you. I’ll have a driver ready to take you out there as soon as you’re ready.”

Well, I missed out on the picnic.

Several said they would see that the little girl got food every day and they would try to tell her where I had gone.

Talavera had learned where his brother was buried and I had planned to go with him to visit the cemetery so I missed out on going there, too.

CHAPTER TWENTYTHREE

It was only a few miles to the crossroads Sergeant Reed had directed us to. The truck driver waited while we went to the first house. A couple and 3 teenage daughters and one younger girl lived there. As near as we could understand (he knew a few words of English) the Americans (OSS) had told the people next door to move out and that some soldiers would be arriving that day. So we went to the other house about a hundred yards away and knocked on the door. No one was there, just as the man had said, so we went in. It looked like the people had just walked out and left everything as it was. They had taken their food and clothing. We unloaded, moved in and the truck driver left.

I sent two guys down to the crossroads to set up a guard post and get a campfire going. They would take four-hour shifts. The crossroads was about a quarter of a mile away. The rest of us ate supper and lay down to sleep as night had fallen.

Four hours later two others went to the crossroads. They would have four hours on and eight hours off. I rotated them so twice a day there would be someone with 12 hours off. It gets very tiresome and boring standing guard, especially from Midnight to daylight. We usually slept most of the 12 hours off. There was hardly any traffic going by the crossroads, a few vehicles and people on bicycles so after a few days we decided just one would be there just for two hours. That way the seven of us could cover 24 hours and lots of time off.

The second day a couple of OSS or G2 came around and said they wanted us to help search the house where the family lived. They said he was a known Nazi. We searched and found some election ballots. The box for Hitler was already crossed when they were printed!

(They couldn't vote for anyone else.) This was about all we found. They took the man in for questioning but brought him back the next day. The family was teed off at us for a couple of days but got over it.

There was no excitement on the guard post, hardly anyone going by to be checked. With only one guy on guard, the rest of us sat around, played cards or talked to the three girls. Some of us were getting pretty good at German and understanding it. We all had German phrase books that had been issued before we left France.

One day the 13 year old wanted me to go with her to get some milk. I figured she was probably going to the next farm about a half-mile away. She did not go that way but headed up the hill in the opposite direction. As we went over the hill I could see another larger place down the road.

As we got closer, I could hear singing, men's voices and what sounded like a lot of them. When we walked into the place it was full of German soldiers of the army we were discharging. This was a store and a beer house.

They quit singing when they saw me and it got so quiet you could hear a pin drop. I was carrying a carbine, a German Luger and a knife.

The girl got her milk and we left. When we were about a quarter of a mile away, they started singing again.

Talk about the hair standing up on the back of your neck.

I never went back there alone, but two of us would walk up there and buy beer after that. It always got quiet when we walked in and would liven up after we had bought our beer and left.

When I found there were so many German soldiers just over the hill, we checked over the hills in every direction and found there were hundreds of them all around us. We went back to two guards on the Crossroads. Never had any trouble out of the German soldiers who had their own officers but there was always the possibility of one or two going berserk.

There was a small stream about 50 yards from the houses and the four girls went there every day to bathe. It wasn't deep enough to swim in. One day they stopped and asked us if we had any soap we could give them. They hadn't had any soap for about seven months. We gave them each a new bar of soap and a couple of guys even offered to help them use it! They understood what the guys were saying. They giggled and said, "Nein, nein." (No, they were not naked down by the stream, they wore bathing suits.)

I believe they liked us pretty well, especially the youngest girl. Whenever any of us were outside she would be right out there, too. Of course, it may have been because we gave her candy and gum once in a while when we had it.

Kesselman and I were sitting at the table eating lunch and here comes the older sister running and yelling, "Corporal, commen sie mitt mere schnell!" (Come with me fast.) "Du haben", I couldn't make out the last words. She pointed at our gun belts and musette bags which both held first aid kits on them.

We grabbed up the belts and bags and followed her. The 13 year old had been wading in the creek and had gotten a terrible cut on her leg, deep, long and jagged. It was bleeding bad. The girls had put rags on it but hadn't helped much. The girl was shaking and crying. She was in a bathing suit and soaking wet. We took one of the large compress type bandages from a kit and wrapped it over the cut. We had always been told not to use a tourniquet if you could keep away from it. I gripped the bandage over the cut real tight and the bleeding slowed.

Two more of our guard mount came up and I had them run back and get a couple of our wool blankets. As soon as they came with the blankets, Kesselman held her wound and I lifted her up and the other two wrapped the blankets around her. It was a warm June day and she soon stopped shaking, which also slowed the bleeding. We gave her a couple of the pills we carried for pain. (They were bigger than aspirin but I don't recall what we called them.)

We said the cut should be washed with warm soapy water and then we could put powdered sulfa on it. When this was done, we put sulfa powder on the cut and re-bandaged it. For the next few days we went to their house and put clean bandages and sulfa on the injury. We used up about all the bandages and sulfa we had and gave them what was left in our kits.

I told some of the guards about a little boy I had seen get killed when I was with Headquarters Battery. He had gotten hit by a truck, cut his jugular vein and even the medics couldn't do anything for him. My first thought when I saw this girl's cut was, "Is this child going to bleed to death?" I guess a cut on the leg isn't as bad as the neck.

(I have a picture of this girl taken a week or so later that shows the bandaged leg.)

This girl would come down to the guard post and sit and talk to me for a couple hours nearly every day. We sat on a big log and she would ask about cowboys, Indians, the cities and the desert. How big was the U.S., where was I from, just hundreds of questions. Many times I couldn't make out what she was asking, she didn't know English but I was getting to know enough German to make a little sense out of a conversation. It was a pleasant way to pass the time on guard, which was usually very boring. I suppose she had a crush on me. It seems the parents must have trusted us to let these girls go for walks with us even late in the evenings.

Hopefully there is a lady in Germany today with a bad scar on her leg who has had a happy life, or maybe she got to come to America as she hoped.

About a month after the little boy with the lacerated jugular bled to death, an American medic took the tube of a fountain pen, put it in a soldier's severed jugular and saved him. This story appeared in the "Stars and Stripes" and nearly everyone read this newspaper. As a consequence, many of the medics started carrying a pen barrel rounded off on the ends.

The German mother even got so she would talk to me but not the dad. He would nod when we spoke to him, and that was all. He seldom came out of the house. We never did figure out what they did for a living. They didn't seem to be the peasant farmer type, except the girls were rather muscular, so maybe they did work in the fields and there was no work that needed to be done at the time we were there.

We were there a little over two weeks. A truck came to pick us up and we said goodbye to the girls and their mother. The man never came out of the house. We shook hands all around. The blond I had gone on walks with, kissed me on the check and the 13 year old started crying. Some of us said we might see her again. Of course we never did. We drove off with them waving.

Rubin Kesselman, Baltimore, Maryland said, "Can you believe how much nicer these people are than the

French and English, and these are the people we have been killing?!"

On the ride back we talked about this being the way it was in all the different places we had been in Germany. We decided the Army of Occupation would not have many problems with the common people.

We got back to the Battery late in the afternoon, unpacked, went to eat and got caught up on the gossip from the other guys. They said the little girl was still coming by every day. They had a softball team organized, were playing different outfits and winning. (They had a pitcher from Denver who was as near a pro as you would find in softball.)

Jim Tynan said it didn't work out between him and the ugly blond DP kitchen girl, but the other gal I was supposed to go on the picnic with was still going out with a GI from Montana that Jim took with him that night. Tynan had seen another blond and a brunette both DPs that were staying at a place a couple miles from the battery and he wanted me to go with him the next day to get a date with them.

Next day the little girl came by again. She was glad to see me and sat down on the bench beside me.

That afternoon Tynan and I went to see if we could find the girls he talked about. We walked about a mile and here they came down the road. The brunette was slim and nice looking. The blond was slim also but not very good looking. I guess Tynan liked ugly blonds.

We talked to them awhile standing there in the middle of the road. He asked them if they wanted to go out walking that evening. They said yes, so he said, "We'll come up to your place to pick you up." At least, that's what he thought he said in German.

But they said, "No, we'll come to your place."

He says, “Help me out, Walsh.”

I said, “You really screwed up on this. We better say goodbye and leave.”

He says, “Why?”

I said, “I think they think they are moving in with us.”

He says, “Oh, shit, surely not.”

I said, “Say goodbye and let’s go.” So we did.

That evening here they came down the street carrying their backpacks, probably all they owned. They walked slowly by the house where we lived. Tynan said, “You were right.” And we ran and hid and told the guys if they stopped and asked about us to tell them we were gone.

The two girls walked by slowly several times until it was almost dark and then left before curfew. The other guys really got on our case.

I blamed it on Tynan. He really screwed up in his German talk! But he thought it was funny now that it was over. Arthur Nickel (Chicago, Illinois), Talavera, John Murray (Sydney, NE), Johnson and some of the other 16 guys in the house wouldn’t let it rest for days. They said they would have been glad to move out of their rooms and let the girls move in. We said, “You know better than that, an officer inspects the houses every day.” They said we could hide them under the bed when we saw the officers coming. We got kidded for days.

The girls were never seen in town again. Actually it is very sad that war and poverty brings people down to this just to get something to eat and have shelter.

Sergeant Reed, Staff Sergeant William Powell (Sheridan, Wyoming), Talavera and I got in on some of the pistol shooting matches that went on. We won some and came close to winning others. Both sergeants used GI Colt 45 cal. I used a 9mm Luger and Talavera a German P38 similar to a Luger except it had a different handgrip that was larger around and made of a shiny brown material. The Luger was plain and always black. Whenever we could, we got Lt. Maddox to go with us and Talavera and I would sit out and let him shoot. These days we always won for he never missed. Once I saw him shoot a running jackrabbit with one shot from a 45 cal. Colt automatic while riding in a moving Jeep.

Lt. Maddox came to A Battery and at his first roll call walked down the rows of men and asked their names. From then on he knew everyone of the 85 or 90 men by name. Everyone liked him. His first name was William so he got the nickname of "Wild Bill"; not to his face, of course. He was the Executive Officer of A Battery and was in charge of all four gun sections of the battery giving them all their commands when firing. You could hear him for a mile when there was a fire mission, getting the men out of their tents and to the Howitzers, yelling range, deflection, etc. He was a real good officer. Some years after the war, at a 242nd reunion, we had a letter from his wife stating that he had died. He had been shot in Dallas, TX. There were rumors among the men at the reunion that his death was mob related. I don't know if it was true.

John Murray, Sidney, Nebraska, was always practicing throwing his knife and got fairly good at it. (During combat he had been with a forward observer party atop a hill. They had left their Jeep in a hurry and the Jeep had been left in neutral. As they took cover to avoid being seen by anyone in the German front line below the hill, the Jeep started rolling slowly downhill toward the Germans! He stuck out his foot and stopped it from rolling any further!) One day he was throwing the knife at a fair sized green tree, hitting the tree point first every time. He took another throw and it hit the tree, butt/handle first, bounced back and the point of the knife struck him in the stomach.

Luckily it went in only about half an inch, and did not pierce any vital organs. Of course he would not go to the doctor. Just put sulfa powder on it and kept it bandaged and taped up.

I never saw him have that knife out of its sheath again.

(Murray came to the 242nd reunion in 1988 and told me he stayed in the Army, was put in the OSS spending a lot of time in Trieste, a hot spot in Europe in the years following the war. He told me he had several close calls there. Trieste had been declared a free port, Americans and British occupied one half of it and the Yugoslavians the other half, but the population was mostly Italian. There was lots of trouble and killing here for almost a decade until the Italians and Yugoslavs made an agreement and the US and British pulled out and turned their area over to the Italians. It was very good to see him again and get to visit with him.)

One evening we were all sitting around talking about what we wanted to eat when we got back to the States. Most wanted steak, but others wanted fried chicken, fish, and so forth. One guy from the East Cost wanted lobster or crab legs. Some wanted a certain vegetable or fruit. Several said they sure would like some shoestring potatoes. (They weren't called "French fries" in 1945.)

When we were on that last guard post, Kesselman and I had figured out how to get some potatoes but at the post we didn't have any lard to cook them in. So we said, "If someone can get some lard from the cooks, or can steal it from the kitchen, we can get some potatoes."

About 6 or 7 of them left, five of them to talk with the cooks and keep them busy while the others went in and stole a whole bunch of lard. Luckily a guy from our house was on guard that night.

In about an hour they were back, “Go get the potatoes.”

Kesselman and I got our musette bags and took off down the road. It was real dark, no moon that night. We knew where a potato field was about three quarters of a mile away. We got down on our stomachs when we got to the field and wiggled our way down between the rows, using our combat knives digging and feeling around under the plants. We got a couple of big potatoes from each plant and patted the dirt back down. It didn’t take long to fill the bags, about 20 pounds of nice big potatoes in each bag. We crawled back to the road and went back to town.

About midnight everyone in the house got a big feed of shoestring potatoes. The cooks missed the lard and thought the DPs probably stole it. They weren’t sure, of course, so everything worked out all right.

Softball games were played about every other day and A Battery was at the top of the league. One day, Corporal Edward DeCesare (New York, New York), the A Battery clerk said, “Let’s get up another team and play those guys a practice game.” So we did. Now DeCesare was a quiet bookish, very smart guy. He wore fairly thick glasses. So he played first base and managed the scrub team and I pitched.

Now the league team had this Denver guy who pitched for them and had real speed, curve, etc. Anyway, we hit him all over the place and managed to get them out every inning and we beat them.

It stopped the bragging we had been hearing about how good they were. There was a lot of poking fun at them for a week or so. DeCesare thought it was really great to shut up that bunch of "smart alecs!"

In this area we were assigned to an artillery group. I believe it was the 402nd F A Group. There would be about anywhere from 4 to 8 or 10 battalions of artillery or about 2000 to 5000 men. This Group's headquarters and the service batteries of the battalions set up a canteen type place in a big building in Rosenhiem which was about 8 miles from where we were in Assling. This canteen was for enlisted men. A smaller building was used for an officers' club.

Each battery could send a few men in a truck to this canteen in the evenings. There were movies every night and German beer to drink and occasionally Coke sent from the States. Now and then hot dogs or hamburgers but no buns. The Army had German bakers baking bread for us using the Army's recipe. There were some German girls doing the serving and cleaning. They were paid and fed.

(There were no young civilian men around this area yet. In fact, the ratio at this time was over 10 women to one man in the age group from about 12 years to 45 years old. After all the prisoners of war were returned, all military discharged and sent home, and even with new male babies being born, the ratio was still five or six to one. Over two generations of men were lost. They had also lost many in World War I, so many that even to the 1990's the ratio hadn't caught up with the pre-World War I ratio.)

When we went to the canteen we would watch movies, play cards, eat and drink. Once a month everyone that had any rank got a liquor ration. This was from corporal on up through the officers. The privates and PFCs didn't get a ration so all the non-coms and officers shared with them. There were some who didn't drink and would give theirs to the privates.

(These rations were not put out in the States, only overseas. In the States liquor could be bought in town. In Oklahoma, which was a "dry" state and only had 3.2 alcohol content beer, liquor could be bought from a bootlegger!)

The ration I usually got was a quart of gin, a quart of scotch, 2 bottles of wine, a large bottle of champagne, sometimes two bottles of champagne, one white, one pink and sometimes a pint of Schnapps, or calvados or brandy. The whiskey, brandy and gin were made in England, Scotland or Ireland. The wine, champagne and calvados came from France, Portugal and Italy.

One time when we got the ration, we picked out privates that were friends to share it with, and a truck load of us went to the club. Everyone else had the same idea so it was a wild night. Some passed out and had to be carried out and put in the truck to go home.

The sergeant who was the head of the number two gun section and I had never got along too good. I never really knew why. I got along good with everyone else. He was a loud mouth from North Dakota and was really kind of dumb. That night at the club, probably because we both had too much to drink, we got into an argument over some little thing. When we were going home in the truck with guys standing up and singing, going down the road 50, 60 miles an hour, this sergeant and I got into a fight, swinging our arms around and nearly fell out the back end of the truck. It was a good thing someone had remembered to fasten the safety strap above the endgate. Somebody grabbed us and pulled us in off the endgate and kept us separated until we got back to our house.

The sergeant passed out just as we were getting into Assling. His section lived in another house and they carried him in.

We had two to carry into our house.

I saw the sergeant the next day and he said, "What was that all about last night?"

I said, "I'll be damned if I know, I guess it was just the booze."

Arthur Nickel (Chicago, Illinois) was standing there and he says, "Well, shake hands and forget it. You're both lucky to be alive. If you would have fallen out of the truck going that speed you wouldn't be around to talk about it today."

There were many interesting places to visit nearby and First Sgt. Reed said that the guy that had done so much guard duty should have the time off first to go to these places. He said, "How about you, Kid Walsh?"

I said, "Sign me up on all of them that come up." He laughed and said okay, and I knew he would.

The next day twelve guys got to go to Salzburg, Austria and I was one of them. With a driver and assistant driver there were 14 of us. Salzburg was less than a hundred miles away and we had two days off. Most of the time the weather was perfect that summer so going in an open truck wasn't bad. Besides we were used to weather after all this time in the Service. Rain, shine, snow, blizzard, I guess we were always ready to go.

In Salzburg we toured castles, famous buildings. Visited a famous opera house. That night we went back to the opera but only stayed about an hour. Didn't any of us care much for it. I believe it was an opera by Wagner.

We went into some really fancy bars. You wouldn't believe the massive oak bars, massive oak tables and chairs and polished oak floors. Must have been places where the big shots hung out.

One castle was built on a hill and made entirely out of wood that was badly weathered. It was really old and crudely built. Even the inside was rough-hewn lumber. We got a kick out of the old guy who was the guide. His English was bad and he kept telling about when times were "dangeroos," and times were more "dangeroos," how the people in the surrounding area came into the castle to be safe from invaders. There was a series of ditches, walls and gates to get behind to defend the castle. Some of the ditches may have been moats but didn't really look like it and the guide didn't mention moats.

(I had always wondered how they got the moats full of water. Several of us talked about this back at Fort Mannonviller in France. That fort was on top of a hill. It would take thousands of people probably years to carry water to fill it and keep it full. The nearest water at Mannonviller was a small stream about a half-mile away down at the bottom of the hill. Of course there would be rainwater runoff from the roof of the castle routed to go in the moat but even in this area of high avereage rainfall it would take a long time. Maybe they didn't keep the moat full all the time, just when there was enough rainfall. I believe we saw about everything of interest in Salzburg. The Army provided a place for us to stay and there was an Army kitchen for meals. Our two days were up and we went back to the battery at Assling.

A couple of days later a load of guys was going to Switzerland. Sergeant Reed said he saved a place for me. Again this wasn't far from Assling. I was amazed at the high mountains, the glaciers, the green valleys and the mountain streams, but most amazed at the cleanliness. City streets were swept and washed every morning, the buildings spic and span inside and out. You wouldn't dream of throwing a cigarette butt on the sidewalk, much less spitting on the sidewalk.

There were real bargains on watches and jewelry. Some men bought watches and jewelry for their wives. Most of us had our own watches plus a military watch issued to us (which would have to be turned back in) and we were not overloaded with occupation marks even though the exchange rate for Swiss marks was good.

The food and beer was good, the cafés clean. The place we stayed at was nice and clean.

One the road back to Assling, a sudden storm came up and it started to hail, and I mean big ones. We were in the open truck, no top to get under. Luck was with us and there was a bridge over the road we were on. We pulled under it and everyone got out and under the truck. It hailed and hailed. This created steam and fog. There was little or no wind so the storm didn't move away from us very fast. Just kept hailing and getting colder. We were about frozen when it finally quit. Hail was a good six inches deep everywhere. The hail stones were all the same size, just a little smaller than ping pong balls.

We drove out from under the bridge, everyone about frozen and this 6x6 truck, even with all wheels driving, could go only about 5 miles an hour through the ice. It was like trying to walk on a bunch of marbles on a slick floor. It took us an hour or more to go 4 or 5 miles. Suddenly we broke out of the steam, the sun was shining and the road was dry.

This was the most hail any of us had ever seen and no one had ever seen hail lay on the level like that. Three of us were from Nebraska and had seen lots of hail but not all the same size laying level.

We made it back to the battery about the time it was getting dark.

Sergeant Reed came around saying there was a Bob Hope USO show in Munich the next night and everyone could go except those on guard. He took down the names of those that wanted to go, about sixty.

The next day was nice and we left for Munich early. We toured around the city, and saw the sights and the damage bombing had done. We stopped and had some beers. We checked out the beer hall where Hitler had held meetings and where the so-called "Beer Hall Putsch" was planned. We saw the statues, the cathedrals and then went to a park I think was in southwest Munich.

A stage was set up at one end of a large open grass area. There was already a large crowd. I took some pictures as we walked down there. The crowd grew to several thousand before the show started. There was Bob Hope, Francis Langford, Jerry Cologna and Les Brown with his "Band of Renown". Also a couple girl singers and a man comedian I hadn't heard of. It was a good show, lasted about an hour and a half.

It was quite a jam getting all the GIs and trucks out of there, but before long we were on our way back to the battery, talking about movie stars and famous people we had seen. There was a cool breeze coming from the south off the Alps. This was something new to us as all the weeks we were in this area there was hardly any wind. We wondered if it was going to storm again but it did not, just stayed real nice.

Edward Walsh, Gresham, Nebraska

Rest and Relaxation

Summer 1945

CHAPTER TWENTYFOUR

We got word there was to be a big show put on by the 402 FA Group. All the battalions were to gather at a big flat pasture. There would be some generals there and all the troops would be in full dress uniform to march and pass in review by a reviewing stand. There would be a band there to lead it.

So we had to clean and check all equipment and clothing and practice marching. The first parade marching we had done since clear back in Camp Howze, Texas.

The day before the big event Talavera and I went up to the house where Sergeant Reed, Sergeant Powell, and DeCesare and the mail clerk Halroyd Brooks (Tipple, West Virginia) and 2 or 3 others stayed. Talavera, Powell and I were the best of friends and had run around together in the States and whenever we had a chance in Europe.

First Sergeant Reed and Sergeant Powell were real regular Army pals because the much older Sergeant Reed had been a buddy of Powell's dad when they were in the horse drawn artillery. Powell's dad had died just before the war. Sergeant Reed was like an uncle to Powell. He and Powell's dad had pulled strings or probably forged papers to get Powell in the Army when he was only 16 years old. He had already been in the regular Army 5 years when I met him. We were about the same age.

Powell and Reed were always going around practicing the fast draw with Colt 45 pistols. They always called each other "Gooney". Some of the guys thought this ranged from disgusting to sick. But not all were aware that the two had known each other since Powell was a baby.

Sergeant Reed said, "Kid, this deal tomorrow is to pin medals on a few guys from each battalion and give out some promotions. They are going to promote Sergeant Major Maloney in Headquarters to Second Lieutenant and promote me to Sergeant Major, both of us to have permanent rank. We are both about ready to retire and it will really help our pensions. This is Colonel Brownlow's, you might say, "gift" to us for being in the Army so long. So I'll be leaving A Battery and moving to Headquarters. I'll sure miss "A" Battery. It's the best outfit I've ever seen in all these years. I never thought I'd say this because the 76th FA was really great. But it's true."

I said, "Let me guess. This "gooney bird" over here is going to be First Sergeant," and I pointed to Powell.

He said, "You're right. They are not going to promote anyone to take Powell's place because there is no regular Army person in the battery. Besides that, we will be going home soon. This point system is the shits. The guys with not enough points will be transferred out, and guys with more points will be transferred in."

(In order to determine who would be discharged from the Army, each man was credited with points, for time in service, for time overseas, wounds endured, battles fought, etc.)

This was to be a step up for Powell who was to stay in the Army for over 30 years. He would get a battlefield commission in Korea for taking over when the officers were all killed. He rapidly went from 2nd Lieutenant Assistant Executive Officer, to First Lieutenant as Executive Officer, to Major in Headquarters, to Lieutenant Colonel as Battalion Commander. He finally spent some time in Vietnam as an advisor and intelligence officer before he retired.

I could see the beginning of the breakup of A Battery.

We had the big parade and the promotions and medals, mostly purple hearts and bronze stars. A lot of the locals gathered around to watch, even some German ex-soldiers. I don't know if any of it impressed them or not. Probably not.

A few days later, First Sergeant Powell said a bunch of guys could go to see Hitler's hideaway. The "Eagle's Nest" was atop a high mountain not too far from where we were. There would be two truckloads of men and one jeep with a driver and officer. Powell was to ride in the back seat of the Jeep and there was room for one more and he asked me.

It looked like it might rain so everyone took raincoats and helmet liners in case it did. And it did. Rained and drizzled all day.

The road going up the mountain was well made, oil matted, but no safety rails or guard rails. I can't stand heights and after we were up several hundred feet I was ready to go back, but of course I didn't say anything about it. I just didn't look off the side or down. The Jeep was finally down to low in four wheel drive. I wondered if it would make it.

We were above the clouds when we finally arrived at the hideaway. We toured the conference room with long tables and fancy chairs. I don't remember how many the table seated. It was easy to imagine Hitler, Goering, von Ribbentrop, Goebels, Borman, and other Nazis and generals sitting around it. We toured the kitchen living quarters, communication center, etc. There was even an elevator to go down in these areas carved out of solid rock down in the top of the mountain.

On the level of the conference room was a big courtyard with a short wall built around it. Below that was a big parking area. On one side of the courtyard you could walk out to a garden and wooded area where the mountainside was not so steep. One side of the courtyard was almost a sheer drop of over 2000 feet or more. Army trucks on the roads below looked like ants when I could see through a break in the clouds below us.

I didn't care much for the drive back down the mountain. This drive was worse than Yarnell Hill in Arizona, but not as bad as Maricopa Pass in California. The mountain, called Ober Salzberg, is near Berchesgarden.

It rained most of the way down but it was warm and we had raincoats and helmet liners to keep dry.

I have checked on a map and some of the mountains in this area are around 6000 feet above sea level.

During this summer a bunch of us got permission to go sight seeing. We went back through here, on down to Innsbruck and crossed over the border into Italy just so we could say we had been in Italy.

A chance came up for a truckload to go to Chiem See (A lake not too far from Assling), Powell picked guys who had always been in his Survey Section and I got to go with them. Talavera, Tynan, Nickel and I and about 10 others with the drivers went there for about a week. We had a beachfront mansion to live in. Must have had twenty rooms, and we had German servants to do the cooking and cleaning. There were boats to use, a sandy beach where we could swim or lay in the sun.

About a thousand yards out in the lake was a fairly large island that had a castle on it. It was called Castle Herrensee (castle on the sea) or Sloss Herrensee Chiemsee. To get there one had to take a larger boat run by Germans. When you got off the boat, there was a wide fancy walk and flower gardens in full bloom and several fountains; really fancy.

The castle was a very large square and must have been about 40 feet tall. The inside just glittered with polished marble floors and walls, and the crystal chandeliers were the biggest I had seen over there. The furniture was massive. Everything had real gold plated trim of some kind on it. Real gold! Some of the walls were covered with tapestry with pure gold and silver threads in it to create the design. The guide told how long it had taken to weave these tapestries by hand. They were several hundred years old. (You may have read about the weavers, slave women, of centuries past and their eye problems.) Some of these tapestries had been done in Europe and others in the middle and Far East. The splendor of the whole thing was unbelievable.

There were glass cases of jewels and coins on display. I would like to have known more about this place, how it came about, who built it, how many workers it took, and how long. One good idea was they did not have to dig and maintain a moat. It had a good thousand yards of lake between it and the shore on all sides.

The island castle and the homes along the shore of the lake - - the whole area was unbelievable. The Germans that owned these places must have been the upper crust in military and manufacturing.

Next day we decided to just lie on the beach and get a tan. We had been going without shirts part of the time playing ball and had a good upper body tan. So we thought we would lie on the beach in our shorts and get our legs tanned.

It worked for most of the guys but I blistered the back of my legs from the hips down and also my inner arms. I could hardly walk back to the house! By the next morning I felt like I had a fever and had a terrific headache. The servants brought food but I didn't want it, just coffee and water. The guys kept stopping in, saying they were going boating, or swimming, or for a walk to the nearby village that had a beer hall and I had better come along. But my legs and head hurt so bad I said, "Go ahead, I'll stay here. Just tell the servants to bring me some water and coffee every couple of hours." I slept most of the day.

About evening Tynan and four or five others came in my room and Tynan said, "We got something to tell you."

I thought, "What the hell did these guys get into now?"

He said they had heard on the radio that the war in the Pacific was probably over, that the Americans had used some new kind of bomb, that one bomb had destroyed a whole big city.

I said, "My God, they have learned how to smash the atom." I had always been an avid reader of the Sunday Omaha Bee News. It was a Hearst newspaper and had carried lots of articles on different ideas on how to smash the atom to create energy and the concern on how to control its chain reaction and so forth. The Hartman Drug Store in Gresham always saved me this paper and also two western magazines for whenever we got into town to pick them up.

Tynan said, "What the hell is an atom?" So I tried to tell them what I had seen in the newspaper and scientists had been afraid they couldn't control the energy and could blow up the world. Some of them didn't believe me or understand a word of what I was talking about.

But Tynan and Nickel told them, "Just wait and see, you'll find out Walsh is right."

I could hardly bend my legs but got up and went to eat. They had brought some beer back with them. Everyone was glad they probably would not have to go to the Pacific theater now. Most of these guys were 19 or 20 and came in as fillers in Fort Sill and Camp Howze and were short quite a few points of getting out of the service. I think I only lacked three points at this time (August) and would have enough time in by the time I got back to the States and had 30 days leave to pick up points. So I hadn't been concerned about it unless they flew us straight to the Pacific but that had not been the policy. It was Europe to the States - one month's furlough, then reassigned to a new outfit and then to the Pacific. For me, there just wouldn't be that much time before I would have the necessary points.

Within a day or two we knew it was an atom bomb. The men wanted to know more about it then, but that was all I knew.

This was a great week at Chiem See even with the sunburn on the legs that ruined about a day and a half of it.

When we were loading up to go back to the battery, another truckload of GIs drove in for a week's stay. We had to talk awhile about the bomb and the younger ones not having to go to the Pacific.

Back at Assling with A Battery there was more ball playing, more guard duty, card playing, over to the canteen for movies, a little local sight seeing, and visiting around in the different houses with the gun sections, wire and radio sections, the First Sergeant's house (now Powell's house) and the drivers' house. The drivers for all sections were supposed to stay with the section they drove for, but early in combat the drivers stole a tent and a stove somewhere and carried them along, and from then on they stayed together. They weren't supposed to, but Captain Carlson let them because he said it dispersed the men around more so it was okay. Here in Assling they got a house of their own.

I had always gone to see them in combat when I would go to the battery from headquarters if I had time. They always wanted the news, if any, and during combat, if they saw me at the battery area, they would wave and holler, "Hey, Rumors, come over and tell us the latest. Joe (coffee) is on!" They made the best Joe in the battery. But Herk Anderson in Radio Section made pretty good coffee too. A little thick - but good.

Everyone knew the day was coming soon for the big split. Over 50% of the men had enough points to go home. The 242nd would head for France by convoy to go to Camp Lucky Strike. There they would turn in equipment and be hauled to a port to go to the States. Those who didn't have enough points would be transferred to another outfit and stay in Germany.

It was a good time – yet it was a sad time. A big percent of us had been together since August of 1942. It was now late August of 1945. About 400 of us had been together since early in 1944. We had several killed, several wounded and sent back to the States. About 10% of our men had been "cannibalized" to be sent to infantry and armor the previous winter (about 50 men) and nearly all of them were killed. (I knew of only two who made it out alive with a scratch and 2 or 3 who were wounded and don't know if they made it or not. I knew of 2 who burned to death in tanks.)

Tynan, Nickel and I tried to explain to the little girl what was going to happen soon, that the Americans were leaving the village and probably no more would live there again and that families could go back to their houses (which I should say, were thoroughly cleaned and left just the way we found them and were also inspected by officers.) We rounded up all the food we could get our hands on and a couple of guys helped her carry it home.

She sat on the benches in the park with us for a couple hours that day. We never knew if she understood what we had said or not.

The day we left a few of the townspeople turned out to see us go. The Burgermeister was around shaking hands with us and the few kids in the village were there for gum and candy. She was there also. She put her arms around my legs. She was crying. I picked her up and held her for a few minutes until I had to get on the truck to leave. Tynan picked her up for a minute and then came to the truck. He was crying. It didn't take much to get him emotional but he wasn't alone this day.

Emotions were flaring all over the place. You wouldn't think this bunch of veterans, many of whom were pretty tough and hardnosed would be so touched. It was very sad.

There were sad goodbyes in the nearby villages too, as the other four batteries were going through the same.

The 242nd saddled up in full force for the last time. Trucks from the 997 FA Battalion filled with men with enough points to go home came in. These men were assigned to the 242nd. This brought the 242nd up to full strength for the first time since the winter before. In other words what the Table of Organization called for – close to 500 men and officers. The rest of us who did not have enough points were then assigned to the 997.

Over the radio came a familiar voice, Colonel Brownlow was speaking, "Close Station! March order! The march order will be Headquarters, Abel, Baker, Charley, Service at close intervals." He gave the cross roads where they would meet and enter the convoy in this order and then "Move 'em out! We're going home!"

This was the last time I ever saw most of them again except for a few at reunions of the 242nd in Kearney,

NE each year. I would miss Talavera and Powell the most and also Sergeant Reed who was at Headquarters now.

The rest of us were now in the 997 Field Artillery Battalion trucks and pulled out behind this long convoy. We gradually picked up other trucks from the 997 who had been at the other four batteries and after a few miles we pulled off on a road headed toward Munich. In a short time we were in the outskirts of the city and in a fancy neighborhood.

I have always wondered how the little German girl got along. However the grain and vegetable crops were good that fall and the Americans were shipping in food for the whole country. General Patton was still in charge of this area and he was doing everything he could to get the Germans working and having them get ready for winter. Also the Marshall Plan would be coming up soon.

William Powell, Sheridan, Wyoming

Edward Walsh, Gresham, Nebraska

Austrian Sightseeing

Summer 1945

Corporal Edward Walsh, Gresham, Nebraska

Assling, Germany

Summer 1945

CHAPTER TWENTYFIVE

When we arrived in Munich we were again assigned houses. The Commanding Officer of the 997 let the former 242nd guys stay together as he did the 997 guys and some from other outfits.

These were two story homes and we were to use the upstairs and leave the downstairs for the family. This family was a man and wife and an 18 year old daughter, whose name was NOT Gretchen. The father was a sculptor. He had a large shop near the house, taller than the two-story house. He was sculpting, carving out of granite, figures that were 40 to 50 feet tall, giant figures of men and women, and one of a giant Nazi eagle with the swastika on it, which was broken up and never hauled away.

That very day a GI came to me and said I was supposed to report to the Battalion Commander. I went with him wondering why I should report to the Battalion Commander, Battery Commander maybe but surely not the Battalion Commander. We went to his office in a house and here in the first room was a Sergeant Major, much, much younger than I was used to in the 242nd. (It flashed through my mind "Sarge you must be a few points short, just like me!")

I said, "I'm Corporal Walsh, I'm supposed to report to the Battalion Commander!" He introduced himself and said, "The Colonel is expecting you. Just knock on the door and go on in and report."

I knocked, opened the door, closed it and stopped about three feet from the desk. I stood at attention and saluted. "Corporal Walsh reporting as ordered, Sir."

He saluted. "At ease. Have a chair."

Hmmmm. This is new. You wouldn't have gotten the "At ease" from Colonel Brownlow and you sure wouldn't have been invited to sit down!

He said, "I see by your files you are an agent and have a driver's license and have a very good record with the 242nd. All of the agents of the 997 had enough points and have gone home. We need an agent to drive back and forth from here to Munich to Third Army Headquarters at Garmish Partenkirchen once a day and sometimes twice. It's a little over 50 miles down there so it could be as much as 200 miles of driving some days. All the Jeep refueling and maintenance will be done at Third Army Headquarters so there may be slight delays when the Jeep needs service. You will need to be in full dress uniform to go there." He hesitated a second. "You probably know what a character General Patton is and how he wants people to dress and act." He laughed.

I said, "Yes, sir."

He said, "Do you want the job?"

I said, "Yes, sir. It sounds like a good experience."

He said, "Good! You start tomorrow. The Sergeant Major will have a Jeep for you and your orders. And also, go fully armed with a carbine and if you have a pistol, take that too."

The next morning I went over to the Headquarters office and there was the Jeep. The Sergeant Major and a Tech Sergeant came out and the Sergeant Major says, "This is Sergeant so and so from the Message Center. He will go with you a couple days to show you around."

I had been a little worried about the first trip, especially when I would get to Third Army Headquarters. I said, "Fine", and we got in the Jeep with a big yellow envelope of papers and took off.

The weather was fine. We drove through villages that had the usual small farm fields around them, and then through foothills, where the forests started and on into more forest and rocky areas. Suddenly we were in the mountains and before long came to Garmish Partenkirchen, a winter sports area and summer vacation spot. There were tall mountains, streams, green valleys and forests, (one of the tallest mountains in the Alps was there, over ten thousand feet), lakes for water sports and fishing in summer, and skating in winter, and steep ski slopes.

The town had a few fancy lodges and a hotel or two. There was a big museum and a large castle like place, something like the one at Chiemsee but not as fancy inside. Third Army Headquarters was in the building and Patton's office was there. He had a large fancy house he was living in.

There was a large area where hundreds of Jeeps and trucks were parked and some buildings where maintenance was done on the vehicles. This was done by German ex-soldiers supervised by American GIs.

The Sergeant introduced me to a German sergeant that talked pretty good English. He was the one that took care of my Jeep every day, checked the oil and everything else, filled it with gas, kept a record on it and serviced it when needed. He washed it every day.

There were probably a hundred or more German guys working around Third Army Headquarters, doing all the work, cooking, cleaning, laundry, cleaning streets, doing mechanic's work and everything that there was any work involved, all supervised by Americans.

The German sergeant seemed as happy as a lark, and sometimes funny. I got to know him pretty well, talking to him every day, sometimes twice a day. I asked him once how much the Americans were paying him and he said very little, but he was getting fed the best he EVER had and the pay was enough to buy cigarettes and beer and toilet articles from the American Army. He was amazed at what all the Americans had over here. He seemed to know how far it had to be shipped from the U.S.

He wasn't married but thought he still had a girl friend. He laughed and said, "Unless she falls for some American."

He wanted to go home and get started on how to immigrate to America. He had picked up a lot of filthy American talk, and I told him he should drop that because American's didn't really talk that way. It was just the service men and they were going to have to stop it when they became civilians again.

He knew I was in the artillery by my uniform and the lettering on the Jeep and he also knew how terrible it was to be under artillery fire. I said, "I know. I've had a few close calls myself."

He goes on, "How did the damn Nazis ever think they could take over the world with the Americans having so much equipment and food they could get over here so fast?"

I said, "The Germans are going to have lots of time to think about that." I told him that we had been instructed when we came into Germany to tell civilians we would be here for at least 20 years or a generation.

He said, "Do you think you really will be here that long?" I said yes I did.

(It's been 50 years now and some Americans are still there.)

I talked to him every day.

I went back and forth usually once a day, sometimes twice a day with messages. The Colonel was trying desperately to get orders to take the 997 home, but was having no luck at all. He wanted me to bring any papers I brought back every day straight to him. He didn't let them go through the Message Center first as procedure dictated. He would go through them and then give them to the Message Center. He was almost frantic about getting orders to go to France and then on to the States. He had been in the Army long enough he should have known you couldn't hurry up ANYTHING!

I didn't do much running around evenings here in Munich, just once in a while to a beer hall with the guys and one USO show, a pretty good show but no stars that we knew. Went to a couple movies somewhere in the area. Played Jackstraws with the family in the house and one of the GIs, Calvin Howard (The Dalles, Oregon) and cards a couple times. The cards looked like tarot cards.

These people were very friendly, and often asked us to come downstairs and play games. The daughter was a pretty girl. The parents kept a close eye on her. No hanky panky with the Americans. They talked different than the peasants in the villages. Guess they were higher class.

I never knew what the rest of the guys in the battalion were doing. I suppose a little guard duty here and there but mostly sitting around waiting. The 242nd guys kind of stayed together and never got acquainted with the ones from the 997 or other outfits who were transferred in. About the only ones I got to know were the Colonel, the Sergeant Major, guys in the Message Center and one Captain I drove for once in awhile in the evening.

These transfer deals were kind of like it had been in combat with the infantry and armor. The officers and non-coms there had gotten so they didn't want to know those transferred in as replacements because they knew they could be dead in a few days. Now that the war was over, everyone seemed to think they had enough friends and didn't care to make any new ones.

Cooler, wet weather set in, cloudy with drizzle off and on but no hard rain. It seemed like the rain would never quit. It was miserable driving an open Jeep in that kind of weather. After about the third day of this, the German sergeant said there were a few Jeeps that had tops out there in those acres and acres of vehicles and that I would get sick driving in all the rain. He said he would have a couple of the German privates go get a top and put it on my Jeep. So they did. They couldn't find one with side curtains but the top helped a lot.

The German sergeant couldn't come up with the word "sick" but I knew what he was trying to say. I said, "Gronk", sick in German and he rattled off a bunch of German and finally got it in English. He wanted to know what "gronk" was in English, and I told him "sick". He said it two or three times. The next day he wanted to know if I got "sick" from being wet. He had learned a new word and it made me feel good that he asked.

By now he would wave as I drove up and was calling me Corporal Walsh. One day he asked if I was German, and I said no I was Irish.

He said, "Irishman. How do you spell Walsh?"

I told him and he said German is Walch or Walsch.

I said, "Oh, is that right." But I already knew this of course.

The cold damp weather was beginning to bother my left knee. I couldn't straighten my left leg driving the Jeep, as the seat was not adjustable. By the time I would get to one end of the trip, I had to limp around awhile to get that leg working again. It was getting worse each day and really got so that I could hardly get out of the Jeep. I didn't want to quit and have nothing to do so I kept on the job. I walked awhile at the end of each trip and took hot baths at night.

The war had been over for about 4 months. I was on a return trip from Third Army, driving through a clearing in the forest when a sniper started shooting at me. The first shot ricocheted off the road under the Jeep. The second hit the Jeep. The third must have gone over my head, or right by me and ricocheted off a boulder on the other side of the road. Three shots was all I was sure of, there could have been more. The clearing was not very wide and I pushed the gas to the floor and got back in the trees.

The sniper was up on the side of the mountain somewhere. He wasn't a very good shot or didn't know how to shoot downhill, but it was too close to suit me. When I got down the road a few miles to a village, I stopped for a few minutes until I quit shaking.

When I got back to the battalion, I went right in to see the Colonel and told him I thought I should have someone ride shotgun with me and he agreed. He had a guard go with me after that and I let him drive some so I could stretch out my leg and it got better.

When we passed through that clearing we would always be going wide open and scrunch down as much as we could. There were never any more shots fired at us. Once was enough to last me for a long time.

My guard was not always the same person every day. Sometimes it might be a 242nd guy, but usually someone from the 997. Tynan, Nickel and Calvin Howard would liked to have had the guard job all the time and I would have liked to have one of them but I didn't give the orders.

I saw General Patton about a block away at the Third Army Headquarters. I had seen him, Eisenhower and Patch during combat about 50 or 60 feet away.

Everything was spit and polish around Third Army Headquarters. Patton was an expert at playing the conqueror. He was a great showman but his mouth always got him in trouble. There were more brass around here, and I mean BIG, than you would ever see at a base in the States. The enlisted men and non-coms were stiff and proper, it's a wonder they didn't break an arm or leg, or back saluting and standing at attention.

I asked the German sergeant at the motor pool what he thought about all this show. He said, "I believe the great General Patton is trying to out-do the Germans and I believe he has already out-done the Germans!" He laughed, "Or Krauts as you call us."

I said, "Now wait a minute. You never heard me call the Germans 'Krauts '. I may say it, but not to a German."

He said, "I know, you are too nice but others do but it's all right. We had a few choice names for you also when the war was on." But he wouldn't tell me what they were. So every day after that I would ask him what he was calling the Americans today. He never told me. After a few days of this, he just laughed and said, "The hell with you." He was a nice guy. I wonder if he ever got to America.

The Colonel was pushing more and more on getting at least to France. The Message Center men said they heard some of the officers say he was going to get in trouble with the Third Army if he didn't ease up a little.

One day word came the unit would convoy to Camp Chesterfield, France, near Reims just northeast of Paris.

So there would be a split of the men again. The younger ones still didn't have enough points. Those with the most points had already gone home but the amount needed now dropped enough that I and the ones who went into the Service when I did would get to go. (This was based on time in service and time overseas only, for those of us who hadn't gotten a medal or purple heart.)

The Colonel pushed the transfers in and out practically day and night and about worked the clerks to death, and our younger buddies from the 242nd had to leave.

It may have been sadder than the first time. Practically everyone cried. Tynan really got broke up, and so did Nickel, Miller, Ness, Overby, Crane and others. I'm not sure how many of the younger ones were transferred out, and where to I don't know.

It took only three days for the Colonel to get this done and on the fourth day we convoyed out of Munich and headed for France. It was early October. I had all of my things in the back of the Jeep I had been driving. The Captain I had driven for a few times brought his stuff and put it in the back and got in with me.

The Colonel and his driver were just ahead of us and he was on the radio. "March Order, Headquarters, Abel, Baker, Charley and Service. Tight intervals." He was giving the crossroads or streets where each battery would fall into the order of march. He was very curt, almost to the point of being nasty. The captain sitting next to me was shaking his head.

The Colonel came over the radio, "I assume you are ready to move? You better be!" Then he said, "Battery Commanders, Report!" Back over the radio came, "Abel Battery, ready to roll", "Baker Battery, ready to roll." "Charley Battery, ready to roll." And last, "Service Battery, ready to roll."

The Colonel was standing up in the truck. "Move them out, we're going to France." He turned and waved for headquarters to move out. His driver started down the road, we pulled in behind him and were on our way.

The convoy looked much the same as the 242nd had looked accept for one thing, the Howitzers were 155mm and quite a bit larger than the 105's the 242nd had used.

As we got started, the Captain said, "When we come to any check points or MP stations, the Colonel will do all the talking and showing of papers. If they say anything to us we are just to say the Colonel has all the papers and knows where we are going."

I almost said why, but I remembered what had happened in basic training and decided to not even try "Why, sir?" So I just said, "Yes, sir." And wondered what was going on.

We didn't quite make it out of Germany the first day and pulled in to a place set up to feed us and some buildings to sleep in, somewhere close to the Rhine River. It was still dark when we got up next morning, had breakfast and were on the road again. My left leg was getting pretty stiff the night before but was pretty good this morning and we would not have as far to go today.

We crossed the Rhine headed toward Riems and Camp Chesterfield. We got there about the middle of the afternoon and stopped just outside of camp, the convoy was strung out along the road.

The Colonel left and went into camp to see where we should go. Everyone dismounted and walked around. I had a hard time getting out of the Jeep, my leg was hurting bad. The Captain had gotten out the other side and didn't see me limping around. After about 15 minutes walking, it got better.

The Colonel was gone about three quarters of an hour and came back telling the Captain, "They don't have a place for us! They say they never got anything from Third Army Headquarters on us. The tent city is completely full." He said he was going back to talk to them again.

When he left, the Captain said to me, "Damn, I knew he was going to get in trouble."

This time I didn't bother to say "Sir", I said, "What are you talking about?"

He said, "I think he badgered Third Army to get those points transferred but they never came through with orders to go to France so he typed up orders to leave Germany himself."

Pretty soon the Colonel came back and said we could pull into a big parking area and could eat at one of the big mess halls but we would have to sleep in or under the trucks for a day or so. We could use the same showers and latrines that the guys in the tent city were using. He didn't seem to be excited about it all.

We got the vehicles grouped up. We then cleaned up and went to the big mess tent and ate. We were outside that night and the following night. An outfit moved out of some of the tents and we moved in.

The tents were pyramidal ones put over boarded up sides like the ones at Yakima and Fort Sill. They slept 8 to 10 guys, but you could get 12 in there.

The third day we turned in the big trucks, trailers and the 155mm Howitzers. We kept a half dozen or so Jeeps for officers to run around on what were supposed to be business trips but was really hanky-panky.

Then the medics showed up. Everyone had to get lined up in rank, in alphabetical order and they came by and shot DDT down the necks of our shirts, up the shirt sleeves, in our pants clear around the belt line, and the pants legs with large powder shooting pumps. It got all over our faces, in our hair and in our boots. We looked like 500 ghosts running around there. Someone said it looked like we got bombed with a planeload of flour. (He had been in the Air Force and had seen planes practice low level bombing with sacks of flour.)

We know about DDT now. But I don't think it ever hurt any of the millions of men around the world that got the same treatment. The Army didn't want fleas, head lice, and body lice, or body crabs brought back to the States. They told us that some lice in Europe lived on rats that carried plague germs. We had all been given plague shots before we went to Europe. I never knew of anyone having lice while in Europe. It's a wonder there wasn't lots of it what with sleeping in barns, sheds, empty buildings, out in the grass, fields and forests.

A few days later we lined up again and got flu shots. There had been some cases in Europe and there was concern there would be an epidemic. Many were sick for several days from the shot. I was really sick for about 3 days. Felt just like I had the flu, high fever and chills, stomach upset. The real flu couldn't have been much worse.

Movies were going on all the time in a giant wood shell of a theater and USO shows about once a week. The USO shows ran several times until everyone that wanted to see them got in.

There was a PX type place where you could buy toilet goods, cigarettes, etc. No beer or liquor here. You had to go on pass to Riems, or Paris to buy it.

Once a month, sometimes every two weeks all the time in Europe we got a free ration of cigarettes, cigars, chewing tobacco both plug and shopped, pipes, pipe tobacco, bags of cigarette tobacco and cigarette papers. Razor blades for the GI razor, shaving cream soap, shave lotion, soap, tooth paste, tooth brush, matches. There was always trading around to get what you really needed. In our battery there was only one guy that smoked cigars so nearly everyone gave him their cigars. Only three chewed tobacco so they got all the chewing tobacco. The few who did not smoke traded the smokes off for something they needed. We found that when you got a German prisoner it was a good ice breaker to give 2 or 3 cigarettes and light one for him. Anyway, the rations and the trading worked real good and everyone was happy with it.

The tent I was in here at Chesterfield was kind of a mixture of guys from the old 242nd. I think there were 5 of us from A Battery, four from headquarters, 2 from service battery and only one from the 997. Out of these, eight were from Nebraska!

Henry Holmes (Taylor, Nebraska) was a real nut. Celeste Holmes, the movie star, was to be at the big theater in a USO show. He was going to try to get by the MPs that did the security around the stage and dressings rooms showing them his ID and claiming he was a relative of hers. He wanted a bunch of us to go with him. Most everyone said no they did not want to be involved in such a nutty plan. We didn't think he would actually do it.

When the show was over, he said, "Okay, let's go meet Celeste!" So we went with him.

He told the MP at the stage door he was a relative of Celeste Holmes, showed his ID and the MP said he would go find out. He comes back and said, "Okay, come with me." We told the MP we were with Holmes and wanted to go too. I think this MP was getting confused about now and he said well he guessed it would be all right.

We went down a hall and there was a civilian guy by her door. The MP stayed. The civilian knocked on the door and out she came. Holmes told her he had the same name and wanted to meet her. She asked where he was from and how long he had been in the Army. She shook his hand and he told her who we were and we just said, "Hi!" He thanked her and she said goodbye and we left.

Going down the hall, the MP said, "You lied to me, didn't you? You knew you were no relation." Holmes said, "Yes, but I got to meet her, didn't I?"

On the way back to the tent, he says, "I should have bet you guys some money I could pull that off.

Holmes came from up in the sand hills of Nebraska. After discharge from the Army, he went to Salt Lake City to find work, married a Mormon girl and joined the church. I saw him once at a Reunion and he had changed so much you wouldn't know he was the same guy.

(Once when I was back in Nebraska, Jeanne and I and her sister Lorene and her husband Clyde Swickard went to a movie in Osceola that starred Celeste Holmes. She was one of Clyde's favorites; he thought she was really sexy. After the movie I told him this story. He was polite about it but I know he didn't believe it. Don't think I ever told it again. Oh, well, you had to be there!)

There was a GI from Chicago in the Headquarters of 997 who had been in B Battery of the 242nd. He was sort of a gangster type and sometime or other had gotten to know some men from the mechanized cavalry who had looted a lot of high priced jewelry from safes, houses and castles during combat. They had always been the first to arrive in a town, the fighting would still be going on and they would blow every safe they came across.

The problem was getting this loot back to the States. They had so much they would never need to work again if they could just get it back to the States. They had cut this Chicago bird in to get some of it back to Chicago, but they didn't get it to him in Munich before we left. After we were in Riems we seldom had roll call. So he said he was going to a nearby air base and catch a ride back to Munich to pick up some of that stuff and would be back in a couple of days. He never came back. He was not missed.

After about a week, word came from Munich that just outside the airbase there had been a shoot-out between five, maybe six GIs and that he was dead and at least 2 others. It seems they got in a fight over some booty. The others had gotten away and still had their ill-gotten gains. Officers came and searched the Chicago's guy's duffle bags and musette bag and went through all his stuff. They found some jewels and quite a bit of money. The officers said it would be sent with his personal belongings to his family in Chicago.

Days were going by and we were just staying there. I figured the Colonel was being penalized for getting us there without orders. We would have to wait until the time we had been supposed to come here before we could move on.

I had driven the same Captain into Riems a couple times. He had never mentioned our situation but one day he told me we would probably be delayed for awhile yet but didn't tell anyone else. I never told anyone else, I didn't think it would be very smart to spread it around.

This Captain started staying at some rich French gal's place overnight. I never found out if he shacked up with her, or if she was running a bawdy house. It was a huge place with an extra cottage for servants.

There was an apartment above the garage for a chauffeur. During the war Germans had taken her cars, so the GIs driving the American officers stayed there. The three times I stayed there, there was another driver (the same one all three times) staying there, too. One night there were three of us and one night four. It was a nice place. We went to the kitchen in the big house to eat with the servants. The food was good – better than in camp. Here was the first and last time I had crepe suzettes or whatever you call them. They looked like flaming pancakes. They were pretty good even if you didn't have white Karo syrup to put on them!

The Captain said I could take the Jeep downtown and take other drivers along. I didn't think this was a very good idea so the one driver who I got to know pretty well and I walked down into the city. It wasn't very far. We toured the bars and clubs and drank some wine. The French beer wasn't very good.

The French were really getting set up to get American dollars. There were many streetwalkers and also gals in the bars out to get a few dollars from the soldiers. Ninety nine per cent of the GIs knew better than to fool around with them. Too many GIs had been beat up, robbed, doped and robbed, and some even killed.

We went back to the apartment before midnight. The nightspots had been really different than the bars and clubs in the States, even different than England and Germany.

October passed and the weather was turning cold and nasty. Not raining or snowing yet but cloudy every day, the wind blowing in from the English Channel and the North Sea. There were beginning to be signs we might move out. Everyone was bored stiff from sitting around, doing nothing, even to the point we wished we were back in Germany doing guard duty. We had turned in our carbines and ammo when we got here and we didn't even have guard duty anymore. Made me feel naked for a few days after carrying them around day and night for so many months. Everyone who wanted to had gone on passes to Riems and Paris and didn't have enough money to go again!

Edward Walsh, Gresham, Nebraska

Assling, Germany

Summer 1945

Preparing to Leave Europe – Fall 1945

Maxwell Stoner, Red Cloud, Nebraska, Unknown, Rudolph, Oeltjen, Fairbury, Nebraska, Henry Holmes, Taylor, Nebraska, Unknown, Camille De Bruyne, Longmont, Colorado, Victor Lutkemier, Campbell, Nebraska, Halroyd Brooks, Tipple, West Virginia, Salvador Talavera, Los Angeles, California, Arnold Johnson, Franklin, Nebraska

CHAPTER TWENTYSIX

The Colonel called us all together, the whole battalion, and announced we were going by train to the south of France to a staging area near Marseilles, that we should wear all the clothes we could get on because the train wasn't heated and was just a shell with wooden seats. He said the cars were old, probably pre-World I or older.

When we packed up this time at least we didn't have to carry a rifle so that helped a little with the weight we had to carry. We were loaded in trucks in March Order, Headquarters, Abel, Baker, Charlie and Service Batteries and went to the train depot. There, we went through the routine of loading by name, rank and serial number as we got on the train.

The Colonel was right. The only place you would find railroad cars like this would have been in a museum or on a movie lot.

One of the ways you could tell the age of the cars was the narrow aisle, low roof, and small seats. We had to bend over to walk down the aisle because GIs were so much taller than the average Frenchman. We could hardly pass each other in the aisle. Four of us were seated in the tiny bench seats facing each other; very uncomfortable, plus we had 2 duffle bags, belt and harness, musette bag and canteen!

We knew that people were much smaller at the turn of the century, but this was ridiculous! At least the engine was halfway modern. Everyone got loaded but it was over two hours before the train started moving. I had problems with my left leg again sitting in the cramped seat in the cold.

The weather was cloudy and gray, windy and every once in a while it would sleet a little. The train was very slow and always stopped on sidings to let other trains by. Several times on the trip it would stop and they would have us get off to walk around and try to get warmed up. When we stopped in the daytime, every kid around would descend upon us begging for gum, chocolate, cigarettes, food or anything we could give them.

We made stops to eat meals at kitchens set up along the route, even a meal at midnight. Of course, all this used up a lot of time.

When we pulled out of Riems, Camp Chesterfield, we headed southwest toward Germany. Everyone was laughing that Hitler must not be dead after all, and had started another war and they were taking us back to Germany to straighten him out. We were traveling through the country where we had been in combat the fall and winter before. We were headed toward Strasburg on the Rhine, going through towns with familiar names, crossing rivers we had to fight to get across the year before.

We actually didn't go into Strasburg but started angling off moving south and finally came down in the rich Rhone Valley and started following the valley and river nearly straight south heading for Marseilles on the Mediterranean Sea.

We pulled in to Lyons about 3 o'clock in the morning for the midnight meal – only 3 hours late. I had been sleeping and woke up about frozen even though I had on heavy wool clothes from long johns on out to the wool overcoat. My mouth was dry from sleeping sitting up. I took a drink out of my canteen but it didn't seem to help. I could hardly get my left leg going. Got off the train and got in the line to eat.

They were serving breakfast by now. The first thing I came to was pineapple juice. I took about 2/3 of a canteen cup full, which is a lot of juice. The juice was cold because the mess tent was out in the open and the weather was cold. It tasted so good I drank it all. Then I took my mess kit and filled it with about 10 strips of bacon, a lot of scrambled powdered eggs and three pieces of toast with real butter. I rinsed out the canteen cup and filled it with hot coffee. The food and coffee were really good.

I got back on the train and went to sit down, luckily by a window. I was suddenly sick, stuck my head out the window and lost my whole breakfast. The pineapple juice wasn't the same coming back up. To this day, I don't care for pineapple juice.

The window was just a piece of glass with a strap on the bottom to pull it up and then hook the strap on a peg above the window to hold it up. That's how old the cars were.

Luckily I was on the side of the car away from the mess line. Felt okay in about a half hour but then I was starved again. I waited a little while and ate a K ration out of my musette bag sitting there in the dark

We had another breakfast stop again about 9:00 AM but this time I laid off the pineapple juice.

It was getting a little warmer the farther south we went but still cloudy and windy. Farther south yet we stopped at a town for a couple hours and the sun was shining and warmer. There was a park across the street from the depot. We went over there and walked around or just sat in the sun. A couple of old French men came around playing accordions and wanting donations. Not money – they mostly wanted cigarettes. Candy and gum were lower on their list and money last. There wasn't anyone too thrilled over the French songs they played.

We finally got to Marseilles. The trip had taken four days. The tracks were not the best, having been patched up since being bombed in many places and the trains traveled slowly, they were playing it safe. These railroads were run by American Railroad Battalions. Engineers, brakemen, firemen, switchmen, all were Americans, even the depot men and the telegraphers. A US train on a good track would have made the trip in 24 hours or less. It's not that far across France from north to south.

The depot in Marseilles was close to the sea and we could see ships out in the harbor. We were put on trucks and carried up winding roads to the top of the seacoast cliffs and the bluffs of the Rhone River valley. Finally got to the top and there were miles of perfectly flat land stretching out like a mesa. The ground looked like solid rock, as if through the years a mountain had been cut off and made level.

Here were rows and rows of tents just like Camp Chesterfield, but probably three times as many. They were set up with one wide street going the full length of the camp, probably 6 or 7 miles with side streets going left and right about every quarter mile or so. These side streets only went about a quarter of a mile long. The camp was long and narrow.

We got to see how long and narrow it was and how many tents because they took us to the far end of the camp and assigned us to tents on the very last street in the camp! Every so many tents had a central shower with washbasins and mirrors but no flush toilets. The toilets were like those in England, board seats with holes. They had dug a deep pit under them like the old outhouses in Nebraska. How they dug these pits in this rocky ground I don't know or how often they had to be covered and a new pit dug. Possible they used chemicals instead of moving the toilets. About a hundred men were using them. I think each toilet had six holes and two troughs for urinals.

Every so far down the long center road was a giant kitchen and mess tent. Each had a permanent staff and fed about 4000 men three times a day. Some French people were also working in these kitchens.

The sun shone nearly every day but it was chilly and the wind blew all the time, kicking up dust on this large flat plateau. Most of the time we stayed in the tents to keep warm, playing cards or reading, or slept, waiting to get the word to get on a ship. There wasn't much else to do. Three or four of us walked the full length of the camp one day. Didn't see any men or outfits we recognized but did find a PX where we could buy a few things.

I was used to walking a lot in combat and also after the war in Germany. There was something I wanted to buy at this PX so I wanted to walk down there again, but couldn't get anyone to go with me. They all said they weren't walking that far. It was at least 4, probably 5 miles there and not much chance of catching a ride. So I went alone.

Walked all the way down there, got what I wanted and started back. Got about a mile when a Jeep went by me with a black driver and a white lieutenant. Suddenly it stopped and backed up. The Lieutenant said, "Corporal, aren't you with the 242nd? Are they here?"

I said I had been with the 242nd but they had gone back to the States and since I didn't have enough points I was with the 997 now.

He said, "Well, I'm in the 999 and we're right across the street from you. We got all switched around with the point system, like everyone else. I recognized you from the road guidepost back at Fort Mannonviller, France when we were in position right beside the 242nd. You remember where it rained so much and the 242nd trucks were always getting stuck and we would pull you guys out with our half tracks?" (The 999 was a black outfit with white officers and they pulled their long barreled 155mm guns with half tracks.)

The 999 and the 242nd were never far apart through most of combat, but usually they were a half mile or more behind us because the 155mm would shoot farther.

He said, "Jump in, we'll give you a ride back." I rode back with them, visited a few minutes, and then the Lieutenant and his driver drove off. I did not see them again.

So we sat and waited – and waited – and waited. Finally the last of November we got word we were leaving. The 997 Colonel that had gotten us out of Germany some illegal way was going to make it home for Christmas. At least he thought so.

But it wasn't to be. At least not for most of us. We had a week to ten days delay because a black guy in the 999 came up missing at roll call. It took a couple days before they finally found him down in the hole of one of the outhouses, throat slit ear to ear

Everyone was confined to tents except to go eat or go to the bathroom. Guards were posted around the entire area and nobody could leave. An investigation started. And of course, some other outfit took our place on the ship and went home.

Wild rumors floated around. Everyone in the 999 was questioned and even some in our 997. One rumor said they had gotten a black GI to confess, and another rumor was that they never did find out who did it. Another rumor was that they had one guy in jail, and another that 2 or 3 guys were in jail. We never did find out for sure. It delayed us over a week!

Back down the steep winding road on trucks to the port at Marseilles. Boarded ship giving name, rank, and serial number. We weren't so far down inside the ship this time. Where we stayed was just above the water line.

This was not a luxury liner. It was a newer troop ship made by Kaiser Shipbuilding on the west coast of the U.S. (This same Kaiser would later make Kaiser and Frasier cars; the auto manufacturing only lasted a few years.) The ship was not very big.

The wind blew hard and the waves on the Mediterranean were high with a steady mist blowing off the tops of the waves. We pulled away from the dock and were on our way. The sea was rougher than any of us had been on before, but this was nothing to what we would get into when we moved out into the Atlantic. I wasn't seasick, but many were.

Next morning at breakfast a sailor said we were moving into the Straits of Gibraltar. No one had said we could not go on deck so a couple of us went up and out on deck to take a look. We could hardly stand up in the wind and the pitching of the ship. But sure enough, off to the right was the Rock of Gibraltar and Spain. We went back in, crossed to the other side of the ship and went out on deck again. Off to this side a little farther away was the coast of Africa extending as far as we could see.

Out into the Atlantic and the sea was getting worse. I was beginning to get a little seasick but still went to eat this day and a couple times the next. Was too sick the next two or three days to eat and my head felt like it would blow up. (We had been issued tickets that were punched each time we ate a meal. I still have the ticket and there are a lot of meals not punched out.)

The waves were so high that when the ship hit them, it went way up high; the nose of the ship would come out of the water and then slam down. It felt like it would break in two in the middle. There was a big compartment in the nose of the ship and this was filled with seawater to hold the nose down. This helped some but it was still really bad. I guess what I called the nose, the sailors called the bow. The ship's newspaper said this was the worst storm in this part of the Atlantic for 40 years. The height of the waves was unbelievable.

After several days we passed through the storm and the sea wasn't quite as rough. We hadn't traveled too far because half the time the ship's propeller or screw was out of the water. It made a terrible noise each time it came out of the water, spun free, and dropped into the water again.

I knew I had to start eating again or be carried off when we got to the States so I forced myself to go eat and had trouble keeping it down at first but finally got all right. In fact I got over the seasickness altogether. Maybe could have been a sailor after all! I was fine by the time we reached port, but there were many that had to be carried off and taken to a hospital to gain some strength back.

Early one morning we neared land and sailed out of the Atlantic Ocean into Chesapeake Bay, through Hampton Roads to Newport News, Virginia. Tugboats pulled us up to the dock and we were unloaded by name, rank, and serial number. There was a band playing but hardly any people to greet the boat. The war had been over nearly four months (since August) so that meeting the troops coming home was about over. And there were hundreds of thousands to come home yet that would not be met.

I never did understand the bitching the Vietnam guys did about this. Several million World War II men came home with no crowds or bands to meet them, as did the ones from the Korean War. I believe most of us liked it that way. We were just glad to get back and go on home. Besides we had done enough marching and really didn't care to march in a parade.

We were taken to Camp Patrick Henry. I believe it was right there by Newport News because I don't remember getting on trucks to drive there. This area was loaded with naval bases, army bases, air bases and forts. (Now there is a NASA base, also.)

We were assigned to a very large barracks and told we could go eat and the mess hall would have anything, ANYTHING, you wanted to eat. Steak, fish, roast, ham chicken, turkey, ANYTHING. All kinds of desserts, ice cream, pop, etc. YOU COULD EVEN ORDER A STEAK COOKED THE WAY YOU WANTED! Unheard of! Also you didn't need a meal ticket, you could go eat anything, any time, 24 hours a day.

This was the biggest mess hall I had ever seen. Seems like it was as big as a football field, and there were lots of lines where you could line up to get food, and hundreds of men eating at the tables. The kitchen help, the servers and waiters were mostly prisoners of war, mostly Italian. They were working here until their turn came to be shipped back to Europe.

Of course, every one ate too much and some got sick, especially those that were not quite over the seasickness yet and had eaten hardly anything aboard ship. Most of us kept going back every few hours when we would think of something else that would be good to eat. We mostly ate and slept!

The second day some of the 997 were put on troop trains going north or south along the coast. Goodbyes were not so sad now.

Then more left for the Chicago area. On the 3rd or 4th day there were enough to fill a train for Fort Leavenworth, Kansas. All of us from the Midwest were put on this train. There were other trains going to the southwest states and the west coast.

We pulled out and before long were winding through the valleys of the so-called mountains of the east coast. We were saying, “Now why do they call these mountains?” After all we were big shots, we had seen the Rockies and the Alps. We really knew what mountains should be!

Everything was pretty well snow covered all the way across Virginia, West Virginia. We crossed the tips of Kentucky, Ohio, Indiana, and Illinois and on to Missouri.

We ate, slept, played cards and talked about the past years and wondered what the future would hold. Things had changed so much in all our lives, no one knew what they would wind up doing. There was not one that was going to stay in the Army. We knew we were going to get a speech at Leavenworth from Recruiting Sergeants trying to get us to enlist in the regular Army, all about the benefits and so forth. Some of these recruiting sergeants got some really nasty answers.

We arrived at a station about six in the morning. I pulled open the window shade of my lower berth (Privates still had to take the upper berths) and the sign on the depot said, “Brookfield, Missouri.” This was where my Aunt Lou (Knouse) Closson lived. I knew where this was!

I said to some of the guys that were up, “How come we are so far north? I thought we were going to Leavenworth, Kansas.” But we kept going west to St. Joseph and somewhere around there crossed the Missouri River into Kansas and then turned south to Leavenworth. The train had made better time than troop trains during the war but it still had taken almost three days to get there from Virginia.

We got to Fort Leavenworth (near Kansas City), loaded onto trucks and were taken to barracks and settled down to wait again. This time to go through the Discharge Center and get out of the Army.

It was a nice barracks and the food was good. Another huge mess hall but not near as big as the one in Camp Patrick Henry, Virginia.

We were given complete physicals. There was lots of paper work, the pitch from the recruiting sergeants, and speeches on going back to civilian life. Lots of promises on taking care of veterans, good separation pay, retraining, never a hospital bill, just go to a VA hospital and any medical problems would be taken care of for the rest of your life, and a promise that there would be a pension if you were eligible. We were told to change our GI life insurance over to any amount we wanted in multiples of one thousand dollars up to ten thousand. Train tickets and meals would be provided for the rest of the way home. If you were taking any souvenir guns home, you must have them registered at your county court house. Also, register your discharge papers and have a billfold size made to carry in your wallet at all times.

We had Christmas dinner, 1945, at Fort Leavenworth. By the 27th of December we said our last goodbyes. Henry Duerr from Staplehurst and I got on the train bound for Lincoln. It was cold, lots of snow everywhere. By the time we got to Union, Nebraska east of Lincoln, it was dark. Here we had to switch trains and get on a local train called a "motor", a small train that ran between Union and Lincoln.

We had to wait an hour or so in Union for this "motor". We got on it and started for Lincoln. After 7 or 8 miles we went over a road crossing that was all iced up. When we hit it, the "motor" almost went off the track. It bounced and lurched so bad it shook people out of their seats, knocked one guy down who was standing up. An old couple slid down between the seats.

This really got Duerr shook and then he got teed off. He went up to the Conductor and told him to slow the damned thing down, that he didn't want to get killed in a damn tin can like this "motor" after making it through the whole damn war. They had already slowed down and did slow down real good for the road crossings for the rest of the way in to Lincoln.

We arrived in Lincoln about 9:30 or 10 PM. Ben Duerr, Henry's dad was waiting. We got into his car and headed for Seward. I never did know what arrangements had been made with Ben, perhaps my Dad or my brother. Ben and my Dad were good friends.

It was about midnight when we stopped in front of Bob's house in Seward (517 South First St.) Bob, Sylvia, Bill age 10, and little Eva Lou were all up waiting for me. They had kept Eva Lou up because she wanted to see who her uncle was. My Dad had gone to bed at his usual eight o'clock and now he came downstairs. (He never seemed to get too excited over anything, like not going to the cave when a tornado was a mile away and coming right toward him.) We talked awhile and then went to bed.

The next day or so we went to Friend to see the Pratts. We hadn't been there ten minutes when Opal wanted to know if I ever killed anyone.

I said, "No."

Then she wanted to know if I ever came close to killing anyone.

I said, "No, not directly like you are thinking, but indirectly, yes. I pulled the lanyard on a 105mm Howitzer many times, and was in a battalion and a battery that worked as a team and killed hundreds of Germans, civilians, livestock, and demolished houses.

When you were up front as a forward observer you could see it happen, otherwise when you were further back at the guns you didn't know what happened until the battalion moved forward and then you saw what you had done as a team. Besides pulling the lanyard, I helped load the Howitzers, carried messages with firing directions for barrages on certain targets. When you saw what had happened, it was rather sickening. But it was us or them – and I got used to it.

The 242nd Field Artillery Battalion and the 997 Field Artillery Battalion were de-activated in December 1945, and passed into the annals of history.

EPILOGUE

I have made an effort to cover the time period with stories I thought were interesting. The buildup and training of a crack outfit like this and its final deadly use. I tried not to put in too many gruesome details

The Official 242nd book does not list all of the killed or all of the wounded. I guess it was hurriedly put together.

My stories deal mostly with only two of the five batterys, A Battery and Headquarters Battery. I was always assigned to A Battery but spent most of the time in combat at Headquarters Battery. There are many stories about here in the States and in Europe but did not have room for all of them.

In a war zone the terrible noise of combat and the results of it are bad, but also the terrible plight of the civilians and refugees and the displaced persons is a sad sight to see. Especially the hungry children. The Germans had caused this to be much worse due to their shipping in hundreds of thousands of people from conquered countries for slave labor. The shortage of food was to continue for some time after combat ended.

However, as tough as General Patton was, he was having the Germans organize the cities and villages in the summer of '45, bringing in powdered milk, flour and other foods and stockpiling it for the coming winter. He was having them cut down trees even the national forests for fuel. The big brass in Europe, the English and the French, even Washington DC, didn't approve of any of this. They wanted the people to endure the hardships of cold and hunger to learn of the aftermath of war. They also wanted Patton out of control of his area because of his loud mouth and his insults of the Russians.

Patton was killed in an auto accident under mysterious conditions. (Died of injuries in this accident.) But he had probably saved many lives with this stockpiling.

General Marshall and President Roosevelt believed like Patton that these people had to be cared for. The Marshall Plan was started to feed and supply food for them. If this had not been done, there is no doubt tens of thousands would have died, especially in the Ruhr valley and other factory areas where there was no room to raise food and no forests to cut down for fuel.

We were told to tell the civilians we would be in Germany at least 20 years with troops and equipment. It is over 50 years now and we are still there. It has been costly but it has kept the German's greed for territorial expansion under control.

Perhaps of all the short phrases used during the war in Europe like: C'est le guerre (it's the war), Le Bosch, Damn Krauts, Hitler Kaput, Mon Deutchland Kaput, How long is this gonna last, Shit, it's time for guard again, I damn near got my ass shot off, - -. The one everyone wanted to hear was French. "Fini le guerre" THE WAR IS OVER.

REUNIONS

In the late 1940s, a group of 242nd veterans from over Nebraska began having a yearly reunion at Harmon Park in Kearney on the 3rd Sunday in August. Over the years, addresses were found for many of the original outfit and the reunion continues to this time, 2006. It is now held in the VFW building in Kearney.

At the 1993 reunion I got a new twist on Sergeant Pfander being killed by a sniper the day after the war ended in Europe.

Thad Krason from Ord told me that it was the last day of combat. He said there was a very large grove of trees and the sniper shot from that grove, hitting the Sergeant in the heart and killing him instantly. Infantry around the area called for white phosphorus shells from the artillery, and with their own mortars turned the grove into a blazing inferno. The sniper came out shooting and they killed him.

This is a completely different story than I heard from several others. At least it's more exciting than the one I told earlier in this book.

After all these years many of the men won't talk about things that happened in combat. They seem to have blacked it out of their memories, or still feel it is too horrible to talk about.

Arnold Johnson related this incident at a reunion:

Arnold was attempting to make an ashtray out of a shell for a souvenir like others were doing. Now, after the base of a 105mm shell was cut off for the main part of the ashtray, a 50 cal. bullet was soldered in the center of it using a welding torch and hot solder. Of course, the powder and bullet was removed, the bullet was put in a machine gun and the little copper firing cap (about the sea of a pea) was fired.

Arnold failed to do this very important last part.

He had the torch going on the base of the bullet with molten lead solder all around it. Suddenly it blew out the little copper cap, which went right by his head and splattered molten lead all over his glasses, face and hands. The hot solder ruined his glasses. He couldn't see very good anyway and it was almost impossible to get new glasses.

Arnold says, "It sure bent up that water can I had the shell laying on - - - - I never did make an ashtray!"

Shell Casing Ashtray – Edward Walsh

Made from one of the first 100 artillery shells

Fired by the 242nd into Germany in December 1944

AFTERWORD

From: Don MacArthur (242nd Field Artillery Battalion Company C)

Date: August 8, 1999

To: Ed Walsh

Hi Ed, will give you a little more on the 242nd, after discharge I was asked to help activate the 186 Inf. Bat. of the 41 Div. Became operations Sgt. and discharged Master Sgt. We had a regular Army Col. As an overseer. During a conversation with him (he had been with a Corp Hdqs in 3rd Army can't remember which one) he said, "Every time I saw where the 242nd was coming into or close to our sector I made a point of being where I could observe & that was something to watch! When they came through those French towns with narrow streets, the speed and roar of those trucks and half the time the howitzers were up on one wheel!! I believe it was a fighting outfit that never should be forgotten! You could always rely on the 242nd for fast response!!" I think this says a lot for the good old 242....Talk to you later. Don MacArthur

100th Division, 203
101st Airborne, 27, 72, 136, 173, 174, 188
103rd Division, 188, 189, 201, 203
244th Field Artillery Group., 78
310 Field Artillery Battalion, 101
314th Infantry Regiment, 102
36th Division, 178, 188
402 FA Group, 263, 270
42nd Rainbow Division, 175, 177, 178
44th Division, 120, 134, 166
71st Division, 189
76th Field Artillery Battalion, 5, 20
79th Division, 101, 102, 104, 108, 111, 120, 190
91st Division, 4, 17, 23, 29
997, 278, 279, 282, 283, 285, 286, 288, 294, 295, 304, 305, 308, 312
999, 114, 304, 305
9th Corps, 4, 17, 45
AberGavenny, South Wales, 91
Anderson, Helmer, 163, 173, 197, 277
Anthony, Charles, 114
Armagost, Doris, 69, 70, 82, 83, 84, 143
Barrantoes, Juan, 171
Battle of the Bulge, 100, 164, 165, 169, 182, 188
Block, James, 143
Block, Richard, 143
Borino, Leno, 136
Brooks, Halroyd, 270, 299
Brown, Eunice (Knouse), 144
Brown, Les, 268
Brownlow, John F., 6, 27, 29, 30, 33, 77, 96, 119, 133, 135, 183, 193, 203, 271, 278, 282
Burton, Allen, 4, 112
Camp Adair, 37
Camp Blackbirds, 91
Camp Chesterfield, 289, 291, 294, 301, 303
Camp Howze, 13, 29, 78, 85, 225, 270, 275
Camp Patrick Henry, 308, 310
Camp Roberts, 5, 10, 42, 214
Camp Shanks, 86, 159, 200
Camp White, 4, 5, 7, 9, 17, 21, 25, 26, 27, 28, 29, 30, 38, 45, 52, 57, 77, 85, 119, 129, 141, 159, 198, 214, 215
Campbell, James, 128, 129, 132, 133, 136, 150, 155, 206, 207, 208
Cannoneer's Hop, 12

Carlson, Robert, 66, 106, 107, 118, 119, 120, 126, 144
Carpenter, Reginald, 120, 138, 183
Centers, Raymond, 128, 132
Closson, Lou (Knouse), 309
Cologna, Jerry, 268
Coufal, Anton, 143
Crosby, Bing, 103
Dachau, 201, 202, 220
Debaetes, Ahlene, 142
DeBruyne, Camille, 162, 299
DeCesare, Edward, 262, 263, 270
Duer, Ben, 311
Duer, Henry, 68, 74, 81, 133, 310, 311
Dunbar, Charles, 74
Eagle's Nest, 272
Eisenhower, Dwight General, 111, 205, 288
Eller, Ernest, 169, 175, 176, 177, 204
Emerton, Herbert, 111, 113, 128, 198, 208
Eppenbaugh, Charley, 142
Eppenbaugh, Earl, 69, 82
Felhafer, Walt, 77
Field Artillery School, 20, 46, 52
Fletcher, Victor, 51, 152
Fort Crook, 1, 3
Fort Leavenworth, 6, 308, 310
Fort Lewis, 4, 6, 45, 72
Fort Logan, 2, 6
Fort Mannonviller, 114, 118, 121, 266, 304
Fort Omaha, 1
Fort Ord, 20
Fort Sill, 5, 13, 20, 28, 29, 45, 46, 47, 49, 50, 51, 52, 56, 58, 63, 64, 67, 71, 77, 78, 101, 129, 215, 224, 225, 226, 275, 292
Fort Snelling, 6
Freas, Joe, 40, 44, 196, 218
Geronimo, 49, 50
Goodrich, Walker R., 5, 6, 198
Grenock, Scotland, 90
Hans, Bud, 143
Hans, Lester, 2, 143
Hinman, Donald, 183
Holmes, Celeste, 294, 295
Holmes, Henry, 41, 294, 299
Hope, Bob, 268
Howard, Calvin, 231, 232, 234, 286, 288
Huff, Bess, 82, 144
Huff, Dave, 82, 144
Huff, Twila, 82
IX Corps Artillery Firing Range, 37
Jackson, Robert, 51, 121, 129, 130, 131, 132, 133, 135, 207, 208
Johnson, Arnold, 73, 299, 315

Johnson, Mauritz, 135
Kesselman, Rubin, 216, 255, 256, 257, 261, 262
Knouse, Ben, 83
Knouse, Bob, 83, 144
Knouse, Geraldine, 83, 84
Knouse, Todd, 144
Landsberg, 219, 222, 227, 233, 234, 235, 236
Langford, Francis, 268
Lazanby, Edward, 199, 200
Lepinsky, 13
LeReux, Paul, 133
Lutkemier, Victor, 218, 299
MacArthur, Don, 317
MacFarland, 9, 23, 41
Maddox, William, 125, 126, 127, 183, 260
Maloney, James, 29, 33, 271
March Order, 61, 122, 167, 168, 177, 290, 300
Marshall Plan, 279, 314
Menze, Harold, 81
Menze, Puzz, 69, 82, 83, 84, 143
Monet, Joseph, 51
Murray, John, 216, 259, 260, 261
Naiphe, 4
Nickel, Arthur, 138, 259, 265, 273, 275, 277, 288, 290
OCS, 5, 46, 48, 56, 72, 78, 176
Oeltjen, Rudolph, 186, 299
Officers' Candidate School, 46
Olson, Glen, 142
Oregon, Medford, 4
Pansanante, 42
Parks, Jean, 143
Patch, Alexander General, 1, 111, 175, 201, 288
Patton, George General, 94, 95, 98, 99, 104, 111, 127, 165, 169, 172, 175, 183, 185, 201, 279, 283, 284, 288, 289, 313, 314
Pfander, Raymond, 4, 214, 315
Pokorney, Rosie, 69, 142
Powell, William, 81, 87, 259, 270, 271, 272, 273, 279, 280
Pratt, Buzz, 246
Pratt, Donna, 246
Pratt, Harry, 70
Pratt, Opal, 311
Pratt, Opal (Modlin), 70
Rauch, Martin, 38
Reed, Bert, 26, 52, 65, 66, 101, 173, 183, 186, 219, 236, 252, 253, 259, 265, 266, 268, 270, 271, 279
Reinmiller, Alysius, 77
Replacement Depot, 150
Riel, Laverne, 52
Ritchie School, 70
Roberts, Dale, 4
Rogers, Ginger, 15

Roosevelt, Franklin
President, 314
RSOP, 72
Salskov, Karl, 82, 158, 159, 183
Schrader, Doris, 70, 143
Seventh Army, 1, 175, 201
Seventh Corps, 3
Sharp, Harold, 172
Sievert, Mary, 70, 142
Smelcer, Oliver, 218
Spotted Wolf, Robert, 244
Stoner, Maxwell, 162, 172, 299
Striz, Anton, 218
Swickard, Clyde, 295
Swickard, Lorene, 295
Talavera, Salvador, 66, 67, 81, 108, 119, 190, 252, 259, 270, 273, 279, 299
Texas 36th Division, 177
Third Army, 1, 104, 175, 201, 283, 284, 288, 289, 291, 292
Tobiason, Albert, 41, 72
Trussell, 23
Tynan, Jim, 245, 246, 250, 251, 252, 258, 259, 273, 275, 277, 278, 288, 290
USS Charles M. Hall, 96
USS Thomas Barry, 86
Utah Beach, 97, 99
Van Housen, Amy (Knouse), 83
Van Zyl, George, 22, 163
Wagner, Maurice, 112, 129, 134, 135, 151, 154, 157, 158, 173, 182, 183, 185, 209, 212
Walsh, Alma, 83, 144
Walsh, Bill, 246, 311
Walsh, Bob, 69, 82, 83, 84, 144, 184, 311
Walsh, Edward Sr., 69, 82, 83, 84, 137, 144, 160, 184, 311
Walsh, Eva (Knouse), 69, 82, 84, 144
Walsh, Eva Lou, 311
Walsh, George, 84
Walsh, Kenny, 94
Walsh, Lee, 83, 144, 184
Walsh, Mary, 143
Walsh, Matt, 84
Walsh, Patrick, 84
Walsh, Ralph, 84
Walsh, Sylvia, 84, 144, 311
Walsh, Teresa, 84
Walsh, Tom, 84
Walsh, Vince, 3
Weaver, Jenny, 69
Wilson, Thomas, 113
Winchester, William, 21
Woodworth, Marvin, 4
Yanez, Eulogio, 20
Zazadil, Adolph, 111, 113, 129, 136, 137
Zetterberg, Gus, 38

Made in United States
Cleveland, OH
05 November 2026

47664328R00181